William C. Lesch, PhD
David E. Rupert, MBA

New Product Screening: A Step-Wise Approach

Pre-publication
REVIEWS,
COMMENTARIES,
EVALUATIONS . . .

"THIS TEXT ADDRESSES, FROM A REAL WORLD POINT OF VIEW, MANY OF THE PROBLEMS OF DIFFERENT CULTURES THAT OCCUR BETWEEN THE TECHNOLOGICAL WORLD AND THE WORLD OF PROMOTION.

Products that are screened using this approach have the benefit of an extensive history of successful product evaluations. The authors address a wide variety of issues that arise in actual product management situations that have rarely been discussed in textbooks."

Charles S. Madden, PhD
Ben H. Williams Professor of Marketing, Hanmaker School of Business, Baylor University, Waco, Texas

More pre-publication
REVIEWS, COMMENTARIES, EVALUATIONS . . .

"Product development is a difficult, complex, and risky endeavor. A variety of factors–some controllable, others uncontrollable–determine the final success or failure of the innovation. Lesch and Rupert integrate these factors in their book and the accompanying computer model that is easy to comprehend and apply to new product screening. This computer model brings the applications of new product screening and evaluation tools to easy reach of product designers, managers, and students.

The book is written recognizing the needs of businesspersons. It contains numerous examples that vividly demonstrate the impact of legal, societal, demand, and marketing mix variables that challenge and stimulate readers toward a more successful end result of successful innovation.

In the contemporary business climate of increased productivity and bottom line performance from all units of the organization, THIS BOOK WILL BE INVALUABLE ACROSS MANUFACTURING AND SERVICE BUSINESSES ALIKE."

Raj Arora, PhD
Schutte Professor of Direct Marketing, University of Missouri-Kansas City

"The audience for this book unquestionably consists of maturing small- and medium-size entrepreneurs–firms which, in heady late bloom, have come to realize that further growth will require at least one more round of successful market offerings. The task is inevitably loaded with risk, hard to conceptualize, and downright scary. It fortunately is well facilitated by the step-wise approach

More pre-publication
REVIEWS, COMMENTARIES, EVALUATIONS . . .

(drawn from Wisconsin's respected PIES II screening model) that Lesch and Rupert here espouse.

New Product Screening: A Step-Wise Approach was obviously written with its target uppermost in the authors' minds. This is not arcane ivory-tower stuff (though it has solid research roots), but is instead a practical, sensible guidebook–a script, really–for structuring the difficult early steps in product development. The examples selected to illustrate each step, taken from actual case histories, go a long way both to reinforce and to clarify the authors' paradigm. MANAGERS AND OTHERS CHARGED WITH GETTING NEW GOODS TO MARKET SHOULD FIND THIS BOOK AN ESPECIALLY INFORMATIVE RESOURCE. AS A BONUS, IT'S ALSO AN EXCELLENT READ."

James E. Hazeltine, DBA
Assistant Professor of Marketing,
Northeastern Illinois University,
Chicago

The Haworth Press, Inc.

NOTES FOR PROFESSIONAL LIBRARIANS AND LIBRARY USERS

This is an original book title published by The Haworth Press, Inc. Unless otherwise noted in specific chapters with attribution, materials in this book have not been previously published elsewhere in any format or language.

CONSERVATION AND PRESERVATION NOTES

All books published by The Haworth Press, Inc. are printed on certified ph neutral, acid free book grade paper. This conforms to recommendations ensuring a book's long life without deterioration.

New Product Screening
A Step-Wise Approach

HAWORTH Marketing Resources
Innovations in Practice & Professional Services
William J. Winston, Senior Editor

New, Recent, and Forthcoming Titles:

Long Term Care Administration: The Management of Institutional and Non-Institutional Components of the Continuum of Care by Ben Abramovice

Cases and Select Readings in Health Care Marketing, edited by Robert E. Sweeney, Robert L. Berl, and William J. Winston

Marketing Planning Guide by Robert E. Stevens, David L. Loudon, and William E. Warren

Marketing for Churches and Ministries by Robert E. Stevens and David L. Loudon

The Clinician's Guide to Managed Mental Health Care by Norman Winegar

Framework for Market-Based Hospital Pricing Decisions by Shahram Heshmat

Professional Services Marketing: Strategy and Tactics by F. G. Crane

A Guide to Preparing Cost-Effective Press Releases by Robert H. Loeffler

How to Create Interest-Evoking, Sales-Inducing, Non-Irritating Advertising by Walter Weir

Market Analysis: Assessing Your Business Opportunities by Robert E. Stevens, Philip K. Sherwood, and J. Paul Dunn

Marketing for Attorneys and Law Firms edited by William J. Winston

Selling Without Confrontation by Jack Greening

Persuasive Advertising for Entrepreneurs and Small Business Owners: How to Create More Effective Sales Messages by Jay P. Granat

Marketing Mental Health Services to Managed Care by Norman Winegar and John L. Bistline

New Product Screening: A Step-Wise Approach by William C. Lesch and David E. Rupert

New Product Screening
A Step-Wise Approach

William C. Lesch, PhD
David E. Rupert, MBA

The Haworth Press
New York • London • Norwood (Australia)

The Haworth Press, Inc., 10 Alice Street, Binghamton, NY 13904-1580

Library of Congress Cataloging-in-Publication Data

Lesch, William Clarence, 1956–
New product screening : a step-wise approach / William C. Lesch, David Rupert.
p. cm.
Includes bibliographical references and index.
ISBN 1-56024-404-6 (alk. paper).
1. New products–Marketing–Evaluation. 2. New products–Evaluation. 3. Product management–Data processing. 4. Decision support systems. I. Rupert, David. II. Title.
HF5415.153.L47 1993
658.5′75–dc20 93-15579
CIP

CONTENTS

ABOUT THE AUTHORS

William C. Lesch, PhD, is Associate Professor of Marketing at Illinois State University and holds a joint appointment at the Norwegian School of Marketing in Oslo where he teaches Advertising Management. He is a consultant to more than three dozen regional and national firms on issues relating to marketing and advertising. He has consulted for the Technology Commercialization Centers in the State of Illinois on new product development. Dr. Lesch has published in a variety of professional journals and serves on several editorial boards.

David E. Rupert, MBA, is Product Manager for Penberthy Inc., in Prophetstown, Illinois. He was Project-Coordinator for Technology Advancement at Illinois State University's Technology Center, and Director of Technology Advancement and Business Relations for Thermo Concepts, Inc., in Bloomington, Illinois. He has five years experience in counseling regional inventors in product evaluation and other aspects of technology transfer and commercialization across a wide spectrum of industries. Mr. Rupert is a member of the Technology Transfer Society.

Preface

Bringing new products into the marketplace can be simultaneously exciting, exhausting, tedious, and fast-paced. The competitiveness of American businesses in both local and global markets depends heavily on the development of new or improved goods, and the popular press is a constant reminder of just how poor Americans are or are not at this process.

This manual and accompanying software are intended to advance the general level of sophistication with which entrepreneurs and businesses of varying sizes approach the new product development sequence. While it is often said that the execution of marketing plans is "everything," it is the planning and behind the scenes revision that ultimately predetermines the outcome of market entry. When this pre-commercialization preparation is incomplete or poorly performed, even the best marketing efforts can result in failure.

Illinois State University's Technology Commercialization Center (ISU-TCC) was formed in 1985 by the Illinois Department of Commerce and Community Affairs. Annually, a portion of its funds were allocated (through 1991) to support the development of new employment opportunities for citizens of Illinois, including sponsoring worthwhile new-product ventures.

In 1986 the ISU-TCC acquired computer software to aid in new product screening and evaluation. The Center sought to assist new businesses in choosing potentially successful products and ideas, while helping to expose flaws in those not worthy of substantial investments. Versions of this manual and accompanying software were developed to help in this process.

Numerous persons have contributed to these efforts. Financial assistance from the ISU-TCC, and administrative support from Jerry Abner, Director, are appreciated. An MBA student, Susan Shipley, aided in case research, writing, and manuscript preparation.

David Doss and Kathyrn Gardner of ISU's Applied Computer Science Department designed the accompanying PIES software at the Center. Valuable insights were gained from marketing research students, and from James Hazeltine, now in the Department of Marketing at Northeastern Illinois University/Chicago. Thanks also to Raj Arora (University of Missouri/Kansas City) for commentary on early drafts.

Finally, our thanks to Dr. Joseph Domitrz, Dean of the College of Business at the University of Wisconsin/Whitewater, for permission to use materials from the *PIES–II Manual for Innovation Evaluation,* coauthored by Gerald G. Udell and Kenneth G. Baker. We hope that our effort is received as providing ready access to concepts that have for many years been out of the reach of numerous interested and motivated entrepreneurs and product managers.

Introduction

Newcomers to new product development are often slow to recognize and, even more frequently, fail to follow any formal procedures for the systematic evaluation of new products or new product concepts. The marketplace is a cruel teacher to many who fail to adequately plan for commercialization, especially for individuals and companies inexperienced in the tasks that lie ahead.

As Buggie (1981) has observed, melding creativity with business acumen is organizationally difficult, time consuming, and laden with interpersonal conflict. To underscore the need for careful nurturing and objective attention to the new product development process, one need only recount the annual failure rate for new products. The most popular (and liberal) estimates put this at greater than 80 percent of those actually introduced, with more conservative figures approaching one-third (Crawford 1979). *Thousands* of new products are introduced each year; probably 5 to 10 percent will survive to age two. This failure to succeed is particularly unfortunate since new product development and new product line acquisition are the most expensive undertakings in which a firm can engage.

This monograph has several purposes, each revolving around the central theme of improving new product screening, and thereby increasing market success, while decreasing the number of losers. The text and software routines are intended for the novice, be that the small business owner, hopeful entrepreneur, or corporate manager with little experience in new product screening.

We begin by describing new product development as currently practiced by successful companies and as taught in business schools. Next, we discuss a detailed model called PIES (Preliminary Innovation Evaluation System) for new product concept screening (Udell and Baker 1980). While we do refer to other models, a modified PIES model is intuitively appealing and useful here because of its extensive real-world history. In one center alone, an early version was used in over 5,000 separate product/concept screens.

ADVANTAGES TO SCREENING

Several benefits result from structuring the approach to new product screening. While the obvious and overriding objective is to improve new product decisions, this method also offers some hidden advantages. First, this screening tool is both comprehensive and systematic, whereas many organizational or individual efforts are likely to be incomplete, addressing but a portion of the factors which impact the success of new products.

Second, by formalizing the process, the likelihood of detecting and exposing subjective organizational or managerial biases is improved. Such informal "group-think" can detract from the quality of the decision. The opportunity to eliminate conflicts before they arise, through the use of a formal system, is an enduring benefit.

Third, expressing and sharing expectations and commitments is likely to improve the product planning process as a whole, as it is during the pre-market stage that the "Pygmalian" effect (a self-fulfilling prophesy) is likely to solidify.

Fourth, the presence of a formal process fosters a greater recognition of the need for quality information to support decision making. Indeed, the most overlooked dimension of any new product decision may be the substance and quantity of information on which it rests.

Finally, adherence to this approach is likely to be educational to users, raising their individual and organizational competencies in product screening and development. The author's own experiences with firms ranging in size from the smallest (less than 100,000 dollars in annual sales) to the largest (more than 100 million dollars) supports the conclusion that size alone does not guarantee systematic quality in the new product screening process. This is a special, acquired skill derived equally from the art and science of business, and tempered through experience and knowledge of individual markets.

As competition in a shrinking global market-place becomes increasingly intense, the need to recognize and practice superlative new product development and screening procedures becomes even more pronounced. We hope following this set of procedures will help.

REFERENCES

Buggie, F. *New Product Development Strategies*. New York: American Management Association, 1981.

Crawford, M. "New Product Failure Rates–Facts and Fallacies." *Research Management*, September 1979, pp. 9-13.

Udell, G. G. and K. Baker. *PIES-II Manual for Innovation Evaluation*. Whitewater, WI: Small Business Development Center, 1980.

Chapter 1

The Context of New Product Screening: The New Product Development Process

INTRODUCTION

The new product development process is best viewed as a series of related and sequential stages. A product begins as an idea. If it survives to commercialization, it then undergoes (hopefully) the familiar product life cycle, and ultimately reaches some stage of decline, precipitating a managerial decision to harvest, or withdraw, the product from the market. This chapter traces these stages, beginning with phases of operations necessary to properly evaluate and groom candidates for commercial launch. Those prospects presenting problems which make them potential "losers" should not pass through these screens, being eliminated before substantial investment costs have accrued.

By the close of this chapter, you should understand some of the complexity of product development procedures and see linkages among stages in the product development cycle. You may also recognize some pitfalls in your own planning or evaluation procedures, identifying areas in need of the most attention.

PRODUCT DEVELOPMENT STAGES

Consideration of new products or new product concepts for market entry is usually viewed as a step-wise process (see for example, McGuire 1972; Pessemier 1982; Hisrich and Peters 1984). While these stages differ somewhat in functional detail depending upon the nature of the intended market (e.g., consumer vs. industrial; see McGuire 1972) and overall product or corporate portfolio (Rabino and Wright 1985; Moore 1987), the general framework is reasonably uniform. Products that may be truly innovative (applying totally new technology to market needs, or that are perceived by the buyer as applying technologies in novel or creative ways) will especially benefit from a progressive framework of coordinated attention. The general sequence is pictured in Figure 1.

The alternative scenario is that of the "me-too" product. In this instance a market has already been established or pioneered by another firm, and what is about to be introduced carries few genuinely new ideas and is perceived as just another brand. Some firms prefer this approach to market development, hoping to minimize pre-market development costs and, at the same time, gain the advantages that accrue from only entering markets where demand has already been proven. Clearly, this approach to product development carries lower research and development investments and represents a less risky marketing strategy. This point of view concentrates more resources on marketing and fewer on product development.

STAGE ONE: Idea Generation

The process begins with idea or concept generation. Typically, organizations employ either a market-driven or product-oriented strategy at this stage. The former focuses on the identification of consumer needs and the development of products that satisfy those needs, while the latter centers more on innovation for innovation's sake (a technology-driven approach).

In a market-directed economy such as ours, some scholars see the product-oriented method as an inherently more risky approach, since rational buyers seldom spend their dollars for products they do not need or for which a strong motivational desire cannot be instilled. Generally, it may take longer for products that are generated purely for the sake of science to diffuse because the additional step of market identification must always be performed, and there is no guarantee that a motivated market (one that recognizes needs for the product) exists or can be identified and developed.

Techniques used to generate new product concepts at this stage range from the relatively unstructured and more creative (free

FIGURE 1. Pre-Commercialization

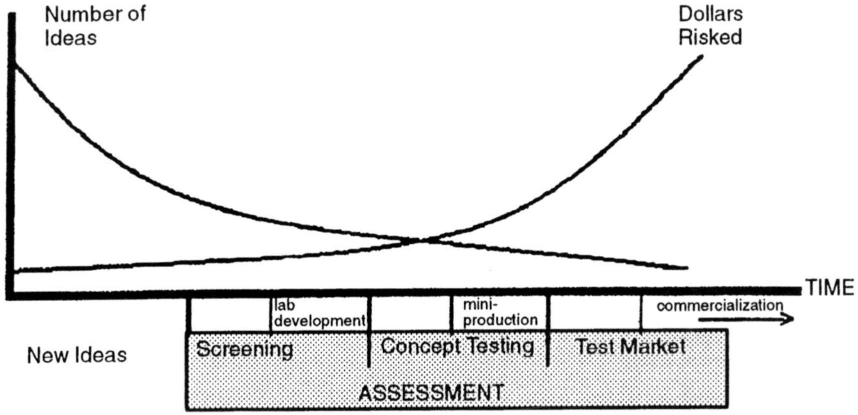

association, brain-storming) to the more structured and focused (reverse engineering, patent searches). See Wind (1982) for detailed insights into these and other consumer and managerial approaches to concept generation.

The purpose of these efforts is to acquire a bank of concepts/ideas from which to draw a subset worthy of evaluation and screening for development. During this phase, however, virtually no judgments with regard to such issues as performance, functional feasibility, and corporate alignment are made. It is usually assumed that large numbers of ideas are required in order to amass an adequate "pool" which will survive through later stages of the evaluation process. Even the kernel of an idea is sometimes of great value later.

Corporate expenditures of time and money on the evaluation of ideas at this preliminary step are not great. This task can be performed by a technical committee which meets infrequently or, depending on the rate of technological change in the marketplace, on a regular basis. While the outcome (whether or not to proceed in development) may be couched in formal terms (detailed checklists, occasional written technical opinions, etc.), the input or level of research buttressing the decisions is not necessarily extensive. Overall, screening out "losers" is probably of greater importance at this point than picking "winners." Too little is known under most circumstances to be able to identify the latter. As such, expert opinions and managerial experience and knowledge are frequently deemed sufficient.

The percentage of prospects discarded at this point is usually quite high. Ames and Hlavacek (1984) suggest a 70 percent rate of loss, while Buggie (1981) reports that during ten years of screening new ideas at the General Foods Corporation, only about 17 percent continued to a level of consideration for product development or, as he referred to it, candidacy.

STAGE TWO: Concept Screening

The procedures employed at the Concept Screening stage vary dramatically, but central issues involve the choice of criteria used in evaluation, and the level of rigor and types of measurement used. Criteria to be used at this stage are often grouped on the basis of similarity, and are chosen as important in some way to future profitability. Shown below are representative categories of factors to be considered at this stage. For other templates, see Wind (1982), O'Meara (1961), Cooper and de Brentani (1984), Kuczmarski (1988), or Crawford (1991). To provide the greatest continuity, Udell and Baker's (1980) list is employed here.

These judgmental criteria or formal screens are, for the most part, self-explanatory but should not be considered exhaustive. Concepts highly rated along these dimensions by definition make good candidates for further corporate investment.

Since an idea or even (later) a prototype seldom receives the

SOCIETAL FACTORS

Legality
Product Safety
Environmental Impact
Societal Impact

DEMAND ANALYSIS FACTORS

Potential Market
Potential Sales

BUSINESS RISK FACTORS

Functional Feasibility
Production Feasibility
Stage of Development
Investment Costs
Payback Period
Profitability
Marketing Research
Research and Development

Trend of Demand
Stability of Demand
Product Life Cycle
Product Line Potential

COMPETITIVE FACTORS

Appearance
Function
Durability
Price
Existing Competition
New Competition
Protection

MARKET ACCEPTANCE FACTORS

Compatibility
Learning
Need
Dependence
Visibility
Promotion
Distribution
Service

highest ratings in all areas, managers must usually establish thresholds or benchmark scores for the various criteria that must be obtained in order for the concept to progress to the next phase, as well as to arrive at an overall assessment. Those that survive receive continued developmental attention.

Remember that every product should to some extent be treated individually for purposes of actual evaluation, as the product-market conditions and the "goodness of fit" of the concept to its intended markets will vary widely. For example, one might have an excellent sugar substitute, but if that market is essentially saturated (with competitors undergoing intense price competition), launch or even further investment may not be wise. Also, depending upon the unique financial condition of the entrepreneur, a decision to proceed will vary. Some individuals are more willing to assume higher levels of risk, frequently weighing the chance of a "hit" against tax factors, for example. Such tax shelter investments (where risk is high and probability of payoff low) are obviously not for everyone. In that case, the payback period may be quite long, and the probability of achieving actual profit objectives low. Further, this serves to underscore the fact that the decision to proceed must address many factors, only a portion of which relate to the purpose or superiority of the product.

The method of measuring the criteria can affect the quality of the judgment as well. As Baker (1980) and Wind (1982) have pointed out, the type of measurement instruments used can have a profound impact on the outcome of the assessment. Generally, capturing more information through the use of an interval-level rating scale is desirable, as opposed to simple nominally-scaled (yes-no) checklists or rankings of product concepts or ideas on each of the criteria. Interval scores usually indicate how much or how little the judge agrees with a particular statement relating to the product, providing more information about the judge's mindset (and degrees of differences among products) than when the judge simply checks a box for approval/disapproval. The PIES scales (see Appendix A) approximate interval-level response data which qualifies them for additional uses. This is particularly important if the firm is interested in testing the overall validity and reliability of the process, since most such tests require interval-level data. Further measurement issues are reserved for discussion in the next chapter.

STAGE THREE: Concept/Early Product Testing

Few ideas survive to this stage, but even among these are some that would probably not thrive in the real marketplace. Risks are now greater, however, since simple managerial judgments may not adequately mirror market acceptance.

To further focus scarce development dollars, and to minimize risks associated with product concepts that predictably will not do well in the marketplace, additional consumer-based tools are used. Focus groups and individual depth interviews are often the least expensive way to gather information from consumers relating to need, product design, appearance, function, and competition. Commonly, the company will redesign various aspects of the products retained, while suspending or terminating the development of concepts that appear to require much more work.

Mail surveys among households that belong to consumer panels can provide a more representative view of a cross-section of the target market, and can aid with the forecasting of likely demand. The rate at which consumers will switch from a competitor's brand, or even convert from a competing product (e.g., use of a mouthwash instead of a toothpaste to fight cavities), can be studied. With this information in hand (and with comparatively little money spent), the marketer can then begin to substantively analyze what might be real opportunities, or actually dismiss all out-of-hand, returning to the drawing board. The opportunity to proceed to the next level is usually very expensive, and justified in only a small percentage of cases.

If the less costly concept tests go well, more advanced tools such as "ASSESSOR" and "ADOPTEST," both heavily statistics-based forecasting methods involving real people in controlled shopping settings, may be used. Usually, a mall interview is used to deliver people to the test store, and the shoppers' behavior under carefully controlled and manipulated conditions is observed and recorded. From these observations predictions are then made (based on the preferences and behavior of the test consumers) about what will occur in real market entry. With some refinement, these can be extremely insightful, coming as close as one market-share percentage point in predictive accuracy. The complexity of these methods makes their use prohibitive, however, except by the more sophisticated and well-heeled manufacturers. Spending 50 to 250,000 dollars would not be unusual. See Wind (1982), Urban and Hauser (1980), or Pessemier (1982) for details.

STAGE FOUR: Test Market

This stage is rapidly developing into a science in our very sophisticated consumer products business. It would not be unusual for a firm to spend 5 million dollars or more evaluating consumer response in smaller, clearly defined markets where the effects of various mixtures of price, product design, promotional levels, and distribution can be considered. Greater insights into consumer behavior are sought, with an eye to what it will take to market the product on a larger scale. Whereas most stage three evaluation can be conducted in a few months, four to six months, or even longer, may be required to perform this fourth stage.

By the close of this phase, as much as two years may have passed since generation of the initial idea. But, as if to complicate things, competitive pressures in some markets are so great that the entire window of opportunity appears to have shrunk to as little as six months. What was once perceived as a managed, stepwise activity conducted at a measured pace is now a flurry of activities best described as "controlled chaos" in the most demanding markets.

Today's computer hardware markets, ever-driven by new technology and intense global competition, are exemplary of this phenomenon.

It should be obvious by now that the process of screening new products and new product concepts can be both complex and intensive. Beginners should be aware that the level of rigor in the process helps to remove as many unexplained reasons for variation in scoring/evaluation/predictions for success as possible. Managers may have honest differences of opinion, but disputes that derive from the methods of evaluation should be minimized. Certainly the latter are within the control of the team or firm and represent a source of potential error in decisions. Further, use of a comprehensive, formal method is evidence of a planned and coordinated approach to the problems associated with screening. The alternative, haphazard or purely subjective assessment, is more likely to result in ruin than any of the present systems because information gathering and attention to detail will have been inadequate. No system is foolproof, but measured, systematic efforts can certainly help to reduce the likelihood or severity of errors.

PRODUCT LIFE CYCLES

The notion of a product life cycle (PLC) is central to the management of products. Simply expressed, products (like all living organisms) can be said to progress, more or less predictably, through a series of stages describing their useful market lives. Beginning with their introduction, products then enter growth, maturity, and decline stages, concluding with market withdrawal. The relevance of these stages to the marketer are briefly described below. Consult Savich and Thompson (1978), Dhalla and Yuspeh (1976), or Wind (1982) for further details. General introductions to the PLC may also be found in Pride and Ferrell (1991) or McCarthy and Perreault (1990). However, keep in mind that the PLC differs in structure when dissimilar product classes, and even similar brands, are being compared. Its usefulness resides in the template that it provides marketing managers, but it should be considered only one of the many descriptive tools at their disposal. The impact of progress through the product life cycle on the financial statement, management focus, etc., is shown in Table 1.

Stages of the Product Life Cycle

During the Introductory phase the manager is most concerned with matters of distribution and promotion, since consumers must be made aware of both the existence of the product and where it can be purchased. Profits are usually low, and the number of competitors (for innovative products) equally small. The general marketing and management effort is directed to establishing a foothold or, as Crawford (1983) termed it, a "beachhead."

The Growth phase is said to be under way when emulators or "clones" developed by competitors begin to appear. Sales rise as new consumers and repeat purchases continue to mount as favorable "word-of-mouth" adds to the effects of early advertising and promotion. Depending upon the pricing strategy employed by the manufacturer, profits can be moderate (due to a high price, skimming strategy) or growing (the result of a low price, penetration strategy).

The strongest earnings can be expected during the Maturity phase. By this time the manufacturer has had the opportunity to take full advantage of the learning curve and reduce the costs associated with production, promotion, and distribution. Consumers are mostly aware of the product, such that advertising expenditures are likely to decline in relationship to sales. Competition has probably focused on product differentiation (based on product extensions and/or modifications), and the number of competitors has peaked.

During the Decline phase, the number of competitors begins to reduce as those with higher unit costs cannot compete during periods of intense price competition and still remain profitable. At a later point, producers frequently attempt to keep the product "alive" with minimal marketing support, intending to withdraw it altogether when revenues dip below the break-even point. Conversely, occasionally manufacturers will attempt to reposition the product in a new market or promote new uses in the existing mar-

TABLE 1. Income Statement (Savich and Thompson 1978)

INCOME STATEMENT

	Start-up	*Growth*	*Maturity*	*Harvest*
Sales	Low but rising	Rising fast	Level	Declining
Cost of Goods Sold				
Direct material	Lots of scrap	Rising	Search for efficiencies	Tied to sales
Direct labor	High per unit	Learning curve benefit	Dependent on union negotiations	Reduced considerably
Overhead				
Depreciation	High due to accelerated depreciation	Rising due to acquisitions	Stable	Low
Indirect labor	Minimal	Growing with labor force	Stable	Redirected
Taxes, property	Minimal	Growing with equipment	Depend on mill levy	Down
Utilities	Related to R&D	Growing with equipment	Stable	Reduced
Gross Profit	Fluctuating	% established	Smooth	Market bearing
Expenses				
Sales Compensation	Variable	Individual rewards	Group rewards	Fixed
Advertising	High	Rising	Stable	None
Travel and entertainment	Very high	Grows with sales force	Reduced	None
Market research	Very high	Leveling	Seek new markets	Accept defeat
General and administrative				
Salaries	Division management	Expanding with work force	Administrative	Reduced
Depreciation	Minimal	Growing with furniture	Stable	None

	Start-up	*Growth*	*Maturity*	*Harvest*
Research and development	Very high	Production oriented	Cost-reduction oriented	Look for new products
Income taxes	Negative	Rising	Stable	Declining
Distribution costs	Looking	Rising	Stable	Cut back
Accounting and information	Start-up costs	Formalizing	Stable	Looking for benefits
Insurance	Minimal	High	Constant and business interruption	Unnecessary
Earnings	Lowest	Good	Highest	Below average
Earnings per Share	Lowest	Average	Highest	Average

kets, striving to revive the product's profitability. Otherwise, as sales continue to decline and the marketer "coasts," a decision to harvest (pull from the market) is inevitable.

It would be expected that some of the profits from the maturity stage would have been directed to the development of other new products for the managed phase-in of replacements that will maintain the company as a meaningful contender in markets of its choice. When the product is finally withdrawn from the marketplace, its life cycle is said to have ended.

As hinted above, marketers frequently "hedge their bets" in the marketplace, particularly by developing and executing long-term product-market strategies. Such "portfolio" management is the key to security for larger firms where overhead costs are great and the number and nature of competitors fierce. Probably the single most important variable predictive of long-term profits for a firm in any given market is market share. Thus, it is critical to the team managing the product that not only launch, but each of the subsequent stages, is managed in such a manner as to ensure a fair share of the available market. Coherent and coordinated product management policies are integral to sound product planning and the long-term health of the firm. Thus, as old products are ushered out, new ones should be in the wings, prepared for launch.

The practical differences and temporal distance among stages are clearly diminishing. Companies are intensifying attention to new product development which results in shorter lines of communication, better coordination, and faster response to market signals. Exemplary of this trend is Deere & Company, a Moline, Illinois-based manufacturer of construction and agricultural equipment. As shown in Figure 2, this compression of activities resulted in delivery of a normally discontinuous innovation product in a period less than half of what had been standard.

In this case, *idea* evaluation and concept evaluation differs from the true "new" product scenario primarily in degree and breadth. Here an existing product was subjected to upgrade. Before prototypes were developed, careful consideration and input was obtained from all of the company's disciplines, including engineering, marketing, manufacturing, purchasing, and testing. In fact, within the context of nearly constant interaction among all corporate elements and with continuous *market* contact, this process evolved not only efficiently, but quite effectively in condensed form. Deere & Co. was able to reduce development costs by ten percent, raise its margins, and enjoy the benefits of rapid market entry. See Blocker (1987) for details.

SUMMARY

The management of products from birth through harvest is a complex and frequently trying ordeal. It is wrought with many pitfalls, some relating to competitive actions and some of the innovator's own making. It is most important that a person new to the process develop a plan to account for and manage the factors that augur success with new products. The axiom "the lack of a plan is a plan for disaster" applies here, as the risks are steep and potentially quite costly. In the next chapter we elaborate upon factors that should be included in such a matrix of screening considerations–those having greatest application in the early product evaluation stage of the new product development cycle.

FIGURE 2. Redesign Process of Major Manufacturer

previous redesign time and new

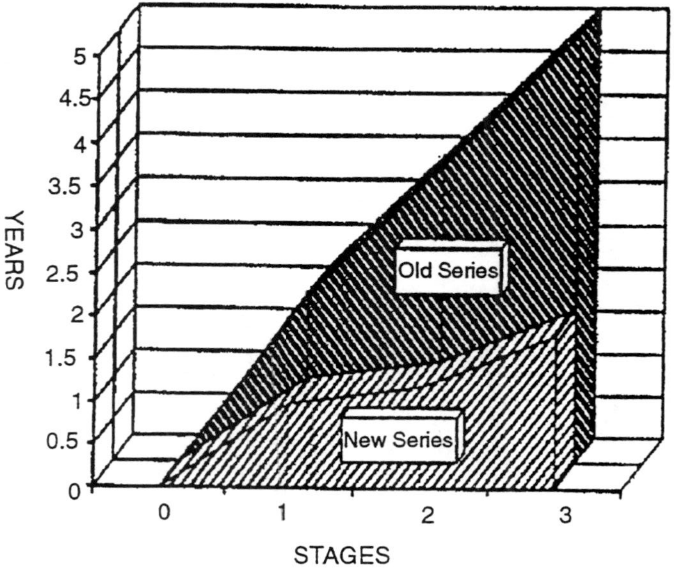

Old Series:

First stage
- feasibility study
- first design
- first build

Second stage
- test
- redesign
- second build

Third stage
- test
- redesign
- pre-production; ship parts to different facility; product support produces literature
- production

New Series:

First stage
- feasibility study
- design: eliminate formal draft and rely on drawings
- try changes on existing machine

Second stage
- first build in different facility
- test

Third stage
- build pre-production machines
- production

REFERENCES

Ames, A. C., and J. Hlavacek. *Managerial Marketing for Industrial Firms.* New York: Random House, 1984.

Baker, K. "The Comparative Analysis of Models for Use in New Product Screening Decisions." PhD Dissertation, University of Oregon, 1980.

Blocker, S. "Deere Workers Challenge System." *Dubuque (IA): Telegraph Herald*, September 20, 1987, p. 86.

Buggie, F. *New Product Development Strategies.* New York: American Management Association, 1981.

Cooper, R. G. and U. de Brentani. "Criteria for Screening New Industrial Products." *Industrial Marketing Management*, Vol. 13, 1984, pp. 149-56.

Crawford, M. *New Products Management.* Homewood, IL: Irwin, Inc., 1983.

________. *New Products Management*, 3rd Ed. Homewood, IL: Irwin, Inc., 1991.

Dhalla, N. K. and S. Yuspeh. "Forget the Product Life Cycle Concept." *Harvard Business Review*, Jan.-Feb. 1976, pp. 102-12.

Hisrich, R. and M. Peters. *Marketing Decisions for New and Mature Products: Planning, Development, and Control.* Columbus, OH: C. Merrill Co., 1984.

Kuczmarski, T. D. *Managing New Products.* Englewood Cliffs, NJ: Prentice Hall, 1988.

McCarthy, E. J. and W. Perreault. *Basic Marketing*, 10th Ed. Homewood, IL: Irwin, Inc., 1990.

McGuire, E. P. *Evaluating New Product Proposals.* New York: The Conference Board, 1972.

Moore, W. L. "New Product Development Practices of Industrial Marketers." *Journal of Product Innovation Management*, Dec. 1987, pp. 6-19.

O'Meara, J. T., Jr. "Selecting Profitable Product." *Harvard Business Review*, Jan.-Feb. 1961, pp. 110-18.

Pessemier, E. *Product Management: Strategy and Organization.* New York: John Wiley and Sons, 1982.

Pride, W. M. and O. C. Ferrell. *Marketing*, 7th Ed. Boston: Houghton Mifflin Company, 1991.

Rabino, S., and A. Wright. "Financial Evaluation of the Product Line." *Journal of Product Innovation Management*, March 1985, pp. 56-65.

Savich, R. S. and L. Thompson. "Resource Allocation Within the Product Life Cycle." *MSU Business Topics*, Autumn 1978.

Udell, G. G. and K. Baker. *PIES-II Manual for Innovation Evaluation.* Whitewater, WI: Small Business Development Center, 1980.

Urban, G. and J. Hauser. *Design and Marketing of New Products.* Englewood Cliffs, NJ: Prentice Hall, 1980.

Wind, Y. *Product Policy: Concepts, Methods, and Strategy.* Reading, MA: Addison-Wesley, 1982.

Chapter 2

An Introduction to Screening Models

INTRODUCTION

Most marketing managers would agree that products and horses have a lot in common: picking winners from losers *before* the race is dicey at best; to stay ahead over the long run, the manager should develop a system for rating the prospects and should be cautious not to expect too much from any single entry.

This successful "stable," or portfolio of products that ultimately develops will provide the mainstay of income for the firm, and should reflect a mixture of opportunities. Some will be in decline, losing share in poor markets, while others will be undergoing rapid growth. Some portion will be in mature, stable condition, generating considerable cash to reinvest elsewhere. Reinvestment should in part be directed to potentially profitable or otherwise successful products and product concepts which are the offspring of careful screening, development, refinement, and test marketing efforts. Thus, an organization implements policies which guide and reflect its goals.

Products in any of the four categories should be periodically reviewed for performance, their objectives reassessed, and their "fit" with the firm's long-term goals evaluated. Thus, by maintaining a stable of "horses" at various stages, an organization reduces its overall risk and improves its likelihood of cashing-in on winners.

In screening the crop of candidates that will receive early grooming and development (only a few of which will ultimately survive to prove themselves in the market), finances are a very important concern. This chapter outlines issues that the manager should address, detailing one model for performing the screening task. By the close of this chapter you should better appreciate the difficulties associated with early development-stage evaluation. Also, your knowledge of one model should be sufficient for you to feel greater confidence in knowing what factors should be employed, as well as how they should be applied. The importance of information gathering prior to decision making will also be made clear.

SCORING MODELS: A BRIEF REVIEW

History is speckled with researchers' attempts to identify the determinants of successful new products. The seminal contributor was O'Meara (1961), with more powerful attempts to isolate specific important dimensions undertaken by Shocker, Gensch, and Simon (1969). Similar efforts at clarification and purification of the measurement tools used in screening, or even to predict success, were proposed by Cooper (1979), Udell and Baker (1980), Calantone and Cooper (1981), Cooper and de Brentani (1984), and Baker and Albaum (1986), to name but a few.

While many firms have developed their own proprietary techniques and data bases, two published lines of research provide the bases for additional research and refinement. The NEWPROD system advocated by Cooper and the PIES methodology are excellent points of departure for tailoring judgment-based screening tools, and represent the most accessible and systematic academic attempts to elucidate a complicated problem. The reader may wish to consult the original sources for additional detail. Below, we discuss issues generic to these and other scoring tools.

ISSUES IN NEW PRODUCT SCREENING MODELS

Validity

Social scientists are always concerned with assessing the validity of their measures. In the case of these models, some attention should be paid to *construct* validity, *domain* validity, and *predictive* validity.

Construct validity asks the question: Is our tool really measuring what it claims to measure? For example, when the evaluator considers the market size criterion, does that judgment accurately reflect market size? Or was there not enough information, or did the rater not understand the meaning behind the term so as to invalidate the response? This is a critical issue in the use of these tools, but one that can be addressed through training and use of good information.

Domain validity tests the issue of whether the instrument adequately captures or represents all aspects of the decision. In other words, if in reality some 35 different factors warranted evaluator attention, and the model we use considers only ten, one could legitimately question domain validity. This concern is directly related to our abilities to sufficiently explain why any concept is or is not worthy of continued support. The scoring model should address all factors that actually need to be addressed. Such a decision involves numerous dimensions. To leave any out could prove disastrous.

Predictive validity addresses the question of how useful the instrument is in actually *predicting* performance. If we were, for example, to track a group of predicted winners, did they succeed? Similarly, do the members of the group classified/rated as "losers" indeed lose in the real world?

Some research has been conducted on all of these issues. The scaling efforts (mentioned above) of Shocker, Gensch, and Simon (1969), Baker (1980), and Calantone and Cooper (1981) were directed to examining construct and domain validity; the studies of Baker and Albaum (1986) and Cooper (1985) addressed predictive validity.

It is clear from this research that some degree of construct validity has been established for at least the two more commonly cited approaches, NEWPROD (Cooper) and PIES (Udell). However, there are disagreements between the two approaches on what items/factors constitute the set of variables worthy of consideration (domain validity). While this may be a factor in the field of testing and application, it raises an interesting question: Can a universal model be developed, one equally valid in all markets? Or should each be fine-tuned to the needs/peculiarities of unique market qualities? Most probably the latter is true.

Finally, on the matter of predictive validity, both Cooper (1985) and Baker and Albaum (1986) demonstrate that skilled use of their tools bears predictive validity. This may exceed 70 percent. However, this should be tempered by the observation that in practice firms will evaluate prospects at different stages of development, resulting in somewhat higher or lower levels of success in prediction. Certainly, we would expect that predictive power would be lowest at the earliest stages, and best at the moment immediately prior to launch. Very little work has been done on this topic, due no doubt in part to the difficulty of obtaining access to corporate "morgues." Keep in mind that even the finest product positioned for a motivated audience may fail if poorly priced, ineptly promoted, or inadequately distributed. Further, a "winner" to Company A may not be so for Company B; expectations of performance and actual support color our ability to adequately test this dimension of instrument performance. For this reason, it is imperative that firms develop their own in-house data bases, reflecting their own markets, goals, and marketing skills.

Reliability

Social scientists are also interested in the degree of consistency or repeatability of a measure. Put this way, how much inter-judge (e.g., if we used three judges) agreement would there be on any or all of the dimensions an instrument measures? Similarly, if we were to repeat the scoring exercise in two or three weeks would we arrive at essentially the same conclusions? Unfortunately, this is an area that has received virtually no published attention. In practice, disagreements are usually resolved by the "huddle" method. Managers should be cautious here: an instrument that proves to be unreliable in practice is likely to be suspect for validity as well. Reliability is certainly a prerequisite of validity, but the converse is not necessarily true.

Factor Weighting

A truly universal model for assessment may not exist. And, given any model, one may want to vary the importance of certain factors, depending upon market dynamics. In the pharmaceutical market, for example, product safety is critical, the payback period always long. The former may be weighted more heavily in assessment, the latter not so much. A flexible instrument allows for such recalibration. Empirical and judgmental inputs should be considered in their assignment. Careful tracking of actual product performance could lead to optimizing weight structures, and improve the predictive performance of the instrument.

Factor Summation

The methods by which individuals or groups sum or add values to arrive at composite outcomes has been given considerable attention by scholars (e.g., Engel, Blackwell, and Kollat 1978; Albers and Brockoff 1980; Stahl and Zimmer 1983; Perreault and Russ 1977). A fine overview of processing can be found in Mowen (1987). Different rules are available, the application of which may lead to radically different outcomes. For a detailed overview of this issue as it relates to scoring models, see Baker and Albaum (1986) or Baker (1980). As adapted from Hawkins, Best, and Coney (1983), the following list shows the major approaches.

DECISION RULES

Conjunctive: Select all (or any, or first) products that surpass a minimum level on *each* relevant evaluative criterion.

Disjunctive: Select all (or any, or first) products that surpass a satisfactory level on *any* relevant evaluative criterion.

Lexicographic: Rank the evaluative criteria in terms of importance. Start with the most important criterion and select the product that scores the highest on that dimension. If two or more products tie, continue through the attributes in order of importance until one of the remaining products outperforms the others.

Compensatory: Select the product that provides the highest total score when the performance ratings for all the relevant attributes are added together for each product.

In a linear compensatory decision mode, the judge follows a rule whereby the outcome is the linear combination of the individual weighted scores applied to each of the criteria. None of the criteria is treated as being any more or less important in contribution to the outcome than any of the others. To illustrate, the "payback period" variable score is just as important as the product "appearance" variable score. In this approach, high scores tend to cancel the effects of low scores, and no single variable is any more important than any other in the decision, since all carry equal weight in the outcome. Hence, the compensatory model should be clear.

In a conjunctive approach to modeling the decision, the judge evaluates the product using cut-off values for all the items used in the test. If the product fails to meet any of the specified cut-off values, the product is dropped from further consideration. In practice, this is a very stringent requirement, as it means that (depending upon where the cut-points are set) the product may have to perform well on even a non-relevant dimension. Failure to meet a standard in any area results in product dismissal, an outcome many managers would agree is too stringent in the majority of early screening situations.

In the disjunctive approach, a path similar to that in the conjunctive setting is followed. The exception, however, is that all products are "allowed" that pass minimal set levels of performance on any variables deemed critical by the firm. It is thus a less stringent screen and will allow more concepts to pass through for additional development or pursuit.

Finally, the lexicographic approach involves criterion-by-criterion evaluation and comparison. Products are compared on each of the items individually until a superior (or inferior) product is found. This, too, is a very stringent mode of evaluation and can become quite complex. For example, products might be examined using subsets of variables rather than individual items. In this case, review is multitiered, and only those products simultaneously performing well on a string of criteria simultaneously will be chosen for further examination.

In a test of difference in outcomes using the alternative models described above, Baker and Albaum (1986) showed that application of each of the models resulted in somewhat different decisions, as might be expected. The choices of "correct" (winners) and "incorrect" (losers) products to submit to additional development varied when the different rules for evaluation were used to predict the status of a set of known winners and losers. Overall, the most conservative model was the conjunctive approach. This is a little-studied area of the new product development process, and additional research will be needed to help managers understand which of the particular models is most appropriate for various product-market contexts.

THE PIES-II MODEL

This method of screening originated at the University of Oregon, which obtained federal monies to develop procedures for screening inventors' ideas. The model was obtained by the University of Wisconsin, and in revised PIES-II form remains in use. (The University of Wisconsin at Whitewater and the University of Wisconsin at Madison offer competitive evaluation tools, the latter touting Cooper's NEWPROD.)

In 1986 Illinois State University developed an agreement with the University of Wisconsin to use the PIES-II methodology, with the intention of improvements and application. The microcomputer software accompanying this text is designed to be self-explanatory, menu-option driven. We recommend that you complete your reading of the manual before moving to the software.

The revised software, PADI, is extremely flexible, allowing the user to set weights, determine decision rules (compensatory or conjunctive only), and vary the degree of report-format detail. Included in the Appendix are data input forms, software instructions, potential applications, and sample output. The original PIES-II criteria have been retained largely for their face validity and relevance to the new-user market. Managers have complete freedom, therefore, to tailor the model to suit the specifics of their markets.

Explicated below are the relevant factors, accompanied by their definitions and examples. Refer to the appendix as you progress through this chapter. These definitions and anchors for the scales have in most cases been altered from the original PIES to improve relevance or enhance scale validity.

Societal Factors

Business exists within society's interests and goals, and activities of the former both affect and are affected by the latter. This section pinpoints dimensions that are important to harmonious relations between business and society as they relate to new products.

Legality

Inquire: Are there laws or administrative rules on the books that might prevent or slow this product from entering into the marketplace?

New products must first withstand the test of legality. In the extreme, society has prohibited the manufacture or sale of some products altogether (e.g., a narcotic such as heroin). Others receive considerable regulatory attention (e.g., morphine or even various antibiotics), while still others seem to escape this issue almost entirely (e.g., fountain pens or kitchen clocks).

Legal considerations include product standard guidelines or restrictions. Performance standards (product functions as intended), safety standards (injury prevention), and material/construction standards (uniform composition of product) are developed by industry associations or agencies to protect the consumer. Compliance with these regulations decreases the potential of being found liable for injuries or damages sustained by users or buyers (Giannetti, et al. 1988).

The Consumer Product Safety Commission was established to protect consumers from possible product defects or hazards. Manufacturers of lawn mowers, for example, are subject to considerable design and component regulation, and a would-be innovator in this product-market must be aware of these before proceeding. The product must conform with these rules, or it cannot be legally marketed. Food products that contain additives will need to establish that the additives are safe for human consumption. At the Federal level the FDA is the regulating agency. Video terminals have been the subject of several investigations by OSHA, another government agency, for eye strain and radiation hazards. Manufacturers are frequently held liable for the effects of their products on individuals and society, and these standards help both parties: manufacturer and user.

For example, personal liability suits have devastated many efforts of entrepreneurs in industry. The most common claims are against the designing engineers. If they failed to use that degree of learning and skill ordinarily exercised by engineers in product design, then they can be liable for damages that may result from the neglect of safety standards.

The failure to put a guard on a stand grinder can be considered negligence on the part of the designer by the courts. An innovator who fails to exercise the same care and caution as any other reasonable person (in the profession) may be found liable for any damages incurred from the faulty design.

Administrative rules apply in other areas also. For example, as part of its disclosure policies, the FDA requires the contents of most foods to be printed on outer packaging; consumers have the right to be informed of the contents.

As an example of the magnitude of the issue, consider the case of fast-food labeling, introduced in early 1986. The FDA wanted people to know that some fast foods contained high fat, sodium, ground chicken skin, sulfur dioxide, or yellow dye No. 5. The fast-food industry claimed that listing all the ingredients on the food packaging would be too expensive and impractical. Furthermore, they argued that customers (in comparing ingredients), would stop patronizing restaurants that used ground chicken skin, jeopardizing the industry (Jacobson 1986).

In another instance, product performance was in question. Searle's Copper 7 brand intrauterine contraceptive device was, from the moment of introduction, in danger of litigation. A. H. Robins' Dalkon Shield had already experienced a 1.1 billion dollar lawsuit expense, driving the company into bankruptcy. Though the FDA had tested and approved Searle's Copper 7 device, the company was still found liable for ill effects (Glaberson 1985). Thus, *even though* a manufacturer met all the guidelines for legal entry, it was still subject to later lawsuits.

Smart manufacturers investigate all likely sources of rules that may affect the marketing of their products. They go further to review industry experiences in consumer satisfaction and safety. They examine their potential exposure to liability suits, and take precautions to inform users and safeguard consumers. If a device does not conform to published or known standards or regulatory requirements, contains potentially harmful ingredients, or does not perform as expected, then a state or federal agency, or even a user, may have the right to litigate. If the producer is found to be negligent or has failed to provide adequate warning of hazards, the results can be disastrous. The question of product safety is detailed next.

Product Safety

Inquire: What are the likely uses of this product and what are the risks to which the user is exposed?
What are some possible unintended uses for the product? Are any hazardous?

Put simply, this category deals with types of hazards presented by the use of the product to the end user. As was clear in the previous section, much of the question of legality of a product for market can hinge upon this question. While related, these issues of market legality and consumer product safety should be treated separately. In the case of lawn mowers, one can market the goods if the standards are met; meeting the standards may not absolve one from liability in a later negligence suit. Also, note that the end user is not necessarily the person who buys. Parents often buy lawn mowers for use by their children; the parents' expertise in the use of the product places them at considerably less risk than that of their children in the same applications.

Virtually all of the states have modeled mini-Consumer Product Safety Commissions (CPSC) after the Federal CPSC. By offering the product for sale, the manufacturer is warranting to the buyer that the product is *safe* for the uses intended. In some cases the courts have gone a bit further and allowed damages even when the product was not used as intended.

The American consumer is protected from hazardous products by various other governmental agencies. Potentially unsafe products must pass certain standards before they can be marketed. No product, however, is likely to be 100 percent safe. Even if a product meets its requirements, hazards may still exist.

"CAUTION: this product may be harmful or fatal." Today's companies are concerned about new product releases because strict liability suits make manufacturers responsible for hazards found in their products.

Asbestos is now recognized as a harmful carcinogen. Exposure to airborne asbestos has led to lung cancer, mesothelioma, and occasionally death among thousands of shipyard workers, miners, and industrial laborers. The impact of asbestos on society has even caused manufacturer bankruptcies. For instance, Manville Company had to declare bankruptcy in 1983 when numerous liability suits and plummeting sales drained it of its working capital. So many, in fact, that Manville alone has set aside some 3.6 billion dollars to address outstanding claims in 1988.

Potential hazards are quite evident in off-road vehicles. Buyer infatuation with these vehicles rose along with sales. The versatility of the vehicles is a definite selling point, but owners can push the jeep-like vehicles too far. *Consumer Reports* tested the steering limits of four off-road vehicles, and alleged a safety hazard arose when maneuvering the Suzuki Samurai.

The Samurai is narrow (51.2 inches from the left to the right wheel), short (80 inches between the front and rear wheels), light (1950 lbs.), and had a somewhat high center of gravity. All these factors, claimed *Consumer Reports*, made the Samurai easier to roll over than similar jeeps. The test performed by *Consumer Reports* included a situation where the driver had to avoid an obstacle at a moderate speed. Fifty feet from the obstacle the test driver had to turn away. At 38 mph the Samurai knocked down the cone and tipped over.

This simulated situation is not an unreasonable one. Sharp maneuvering at 40 mph is a likely occurrence, since this is the speed around some residential areas where children may be playing. The Isuzu Trooper, the Jeep Wrangler, and the Jeep Cherokee may have occasionally knocked over cones at 40 mph, but all these vehicles remained stable (*Consumer Reports* July, 1988).

The National Highway Traffic and Safety Administration had received 44 reports of the Samurai rolling over with 16 fatalities resulting. These statistics were alarming, since the Samurai had only been on the road about a year. A defect investigation and a recall of the vehicles was suggested by both U.S. Senator Timothy Wirth and *Consumer Reports*. The petition was denied. A warning sticker telling of the possible maneuvering problems is already on the vehicle, but concerned parties protested that stickers are not an excuse for allegedly poor design (*Consumer Reports* August, 1988).

The consumer should also be informed that a number of safety regulations that cars must meet do not apply to the sport/utility vehicles. Crush-resistant roofs and sides, head restraints for front passengers, eye-level brake lights, and automatic safety belt systems are not requirements for off-road vehicles.

In other instances, products have been known to cause harm to children. Safety is a primary concern when developing children's toys, because children frequently choke on small detachable parts. Fisher-Price and Avon have had to cope with these issues by careful and continuous modifications of their products.

Quaker Oats Co.'s Fisher-Price division was the subject of a suit concerning "Little People" toy figures. Young Ian Cunningham choked on one of the devices, suffering brain damage from the incident. The family claimed that the product was unsafe, and sued for damages, winning 2.3 million dollars in an out-of-court settlement. Fisher-Price, however, opted to continue manufacturing the toy, citing the incident rate as only four occurrences in the past 20 years (*The Wall Street Journal* May, 1988).

Avon avoided possible lawsuits by recalling its Bumble Babee Plush toys. Apparently, the antennae could be pulled off and swallowed. Three incidents were reported to the company, but no injuries had occurred. When recalling its 350,000 Bumble Babee toys Avon offered not only a refund to its purchasers, but a five dollar gift certificate, the latter to reduce any ill-will (*The Wall Street Journal* November, 1987).

Safety of the consumer should be a primary factor in designing a product. A company's demise could be brought about by legal action from an injured consumer, or even as a result of negative publicity. It cannot be ignored.

Environmental Impact

Inquire: Are there any known or foreseeable environmental side-effects that arise out of the production, distribution, consumption, and/or disposal of the product that may impede further development? Conversely, are there any known or foreseeable direct effects or side effects that would be of benefit to society?

What are the long- and short-term implications to the environment that may arise out of production, distribution, consumption, or disposal of this product? Are any of these unusually beneficial or detrimental? Once again, nearly every state has some form of environmental protection agency, modeled after the Federal EPA.

In addition, numerous public interest groups evaluate new products for potential hazards to the environment. Both consumers and governmental bodies recognize possible environmental problems. For example, some consumers will refuse to buy soft drinks in polystyrene bottles because they feel the limited biodegradability is a threat to the environment. Also, the EPA has recently tightened emissions standards for automobiles, affecting the formulation of gasolines, such that lead and olefin content does not exceed so many parts per million. The greater restrictions were imposed in order to reduce harmful emissions dangerous to the environment.

Litigation of environmental hazard responsibility has become a profitable business for lawyers, inspectors, and environmental consulting companies. If environmental contamination is discovered on private property, current and previous owners, sellers, lenders, and developers may be liable for clean-up costs and subsequent health claims. The party most able to pay is, in many cases, stuck with the bill (Neal 1987).

It is estimated that 100 billion dollars will be spent on asbestos removal over the next 25 years. Ironically, asbestos was considered the "wonder fiber" of the twentieth century. It is a cheap, fire-resistant material. At just 25 cents a square foot, asbestos was used to insulate pipes, ducts, and boilers, fire-proof containers, and even to line car brakes. Unfortunately, it can cost 25 dollars per square foot to remove from a building.

It seems that the property owners cannot win. Building occupants and the EPA are attacking companies from both sides. The "ideal" investment of the 1930s and 1940s has financially devastated corporations with large buildings that need to have asbestos removed. The well-being of society will be preserved, but at the cost of numerous bankruptcies (Richman 1988).

Societal Impact

Inquire: To what degree will this innovation solve a social problem or raise the quality of life in society?

When the telephone was invented, society benefited immediately and dramatically. Does a "Pet Rock" offer the same level of benefit to society? How about the mini-skirt? One should consider the relative advantages to society (improvements in quality of life) when applying this criterion. Innovations of little consequence to society would certainly include crunchier potato chips, electronic calendars for the home, plastic key chains, etc. More substantial innovations of great consequence to society include microcomputers, the compass, the calendar, the typewriter, the automobile, and the airplane.

An invention often provokes a societal response, be it perceived as favorable or not. Who was big enough in the 1980s to challenge the behemoth of fast-food chains, McDonalds? Can the "Burger Barons" defy anyone or anything that confronts their "golden" image? Opponents say, "No, they can't!" The battleground was styrofoam.

Styrofoam affords a light weight, sturdy, economical means of packaging, used as an insulator to keep hot things hot and cool things cool. It appeared to be the savior of the fast-food industry, and it should be no surprise that McDonalds would be the largest user of these products. However, it was not foreseen that the "puffed up" plastic would over-burden landfills and pollute the air.

A coalition of grassroots activists, environmental lobbyists, public interest researchers, and politicians vigorously protested McDonalds' use of styrofoam. The first attack came from Vermonters Organized for Clean-up (VOC). After five days of opponent campaigning, McDonalds' agreed to phase out its use of CFCs, ozone damaging chemicals, by 1990. The second battle was Citizen's Clearinghouse for Hazardous Wastes (CCHW) versus the "Arch" enemy, McDonalds. Community groups armed with fliers and posters vehemently protested the use of styrofoam packaging. Local media grabbed hold of the issue and the protest was even mentioned on Johnny Carson's popular *Tonight Show.* The activist groups could defend their protest actions by pointing out that there is a simple substitute for nonbiodegradable styrofoam: paper.

Society's negative reactions to a technological breakthrough can be threatening to companies, especially when the coalition grows large. In this case, McDonalds was forced to stop its boundless use of styrofoam. Customers were willing to trade the convenience (not of great social benefit) of styrofoam for less pollution (Stults 1987).

Business Risk Factors

This section covers factors associated with managerial risk. Persons involved in the innovation process are, of course, well-acquainted with risk. Many have invested hundreds or thousands of personal hours in the development of an idea, and some may even have invested many thousands of dollars.

Functional Feasibility

Inquire: Will the product perform as intended?

This criterion brings to mind the notion of the perpetual motion machine. The argument against such a machine is, obviously, that it (theoretically) will not work. This category applies to the ability of the concept or innovation to perform its intended purpose. Keep in mind that there are often many purposes for a product in its various forms. For example, when 3M was in the process of developing its "Post-its," engineers initially had no idea that the adhesive would be used for the removable note function. Thus, it is important to consider both the primary applications of the innovation, and whether any side benefits/applications would derive that may also be of functional value.

To be among the world's leading innovators, American companies must continue to investigate new methods of manufacturing. Robotics are among the latest tools for improving production efficiency. However, before purchasing such equipment, companies must decide if it is functionally feasible to implement robots in their manufacturing process. If the robot is not capable of performing the intended task, such investments would not be justified. The question is, "Will it work?"

The industrial robot is no longer merely a device that repeatedly performs a fixed sequence of movements. Robot capabilities have advanced significantly in the past decade, with the latest generations able to "see" (light reflection levels) and touch (heat, dimension, and contact). It is not uncommon to observe robots performing tasks previously done by humans.

Originally, robots were simple machines. Early generations could grasp prepositioned objects and transport them to a predetermined place. Engineers were not satisfied with this operation, however. They envisioned robots as "thinking" machines.

The important question was whether it was functionally feasible to create a robot that could make decisions. Engineers needed to instill human-like qualities in the logical, binary minds of robots.

The sense of touch is a very human quality. Some robots can now measure different dimensions by using mechanical grips. When the vise on the end of a robot arm grasps an object, the dimension of the object can be calculated by the degrees of difference between the contracted vise angle and its original angle. The computer program controlling the robot has an internal loop that rejects an object that is larger than a predetermined dimension. In essence, the robot could sort large objects from small ones. Certainly, this crucial sense of touch has now been achieved mechanically.

One could argue that in order for a robot to replace humans, it requires sight to grasp objects that are not in a predetermined position. In today's machines, infrared light readers installed in the robot arm's vise can provide sight. White dots are painted on the target objects' sides, and when the infrared light bounces off, sensors recognize the increased light intensity. The machine's computer program triggers a "grasp" subroutine when light intensity reaches a predetermined level (Crocker and Guelker 1988). A possible drawback to using the light sensors as eyes is that the working environment must be kept clean. The dots on the boxes cannot be smudged or dirty, or the sensor will fail to recognize the change in light reflection.

Giving robots human characteristics commonly leads to the replacement of humans in some operations. The robots can be used to transfer materials, test the quality of products, and undertake other burdensome tasks previously done by human workers. Capital investment in robots is large, but may be recovered in many cases within a few years. When the robots perform the monotonous tasks, the efficiency level of operations will often improve and the human resources can be applied in other areas of operations.

The functional feasibility of robots in an increasing number of settings is thus extremely high. In many instances not only will future technology add to robotic senses, but it will increase the number of fields of use. It may not be unreasonable for robots to invade American homes to perform, perhaps, the monotonous task of housecleaning. We can only hope.

Production Feasibility

Inquire: Can the product be manufactured at a level of production that satisfies quality and volume requirements?
What resources (personnel and manufacturing) will be required?

This next question is dependent upon the level of advancement of production technology. You will frequently have to rely on consultations with experts in order to determine whether production is possible. Firms are sometimes faced with a golden opportunity only to discover that technology is not yet sufficiently advanced for mass production. Metallurgical engineers, for instance, tell us that it is possible to make gold from lead, but the process is too costly and complex to be of any commercial value.

The production feasibility criterion is applied by assessing the difficulties that may be encountered during the production process. Careful analysis should be performed in the planning stages to reduce the risk of production-related obstacles. For example, if a company does not have the required technology to produce the part, it may not want to invest large sums of money in equipment to acquire the ability.

Production feasibility will need to be justified in cost/benefit analysis, but one must also consider the production process itself. A good example exists in the efforts of Victor Papanek who, in 1966, analyzed the feasibility of producing an inexpensive off-road vehicle for third-world nations.

Papanek hoped to develop a vehicle that could be used by underdeveloped nations in times of disaster, because helicopters were prohibitively expensive. The first task was to find a suitable material with which to make the vehicle body. To satisfy this need, the professor and his graduating class developed "fibergrass." This hybrid of native grasses and fiberglass catalysts produced a lightweight, sturdy casing.

Mechanical production was to be performed in various countries to minimize costs. Heavy-metal work was to be done in Egypt and Libya, Central Africa, India, and Brazil. Electronic ignitions were to be produced in Taiwan, Japan, Puerto Rico, and Liberia. Precision metal work and the power train were to be done in the Chinese Democratic Republic, Indonesia, Ecuador, and Ghana. The fibergrass bodies could be made in local villages.

Several prototypes were built to determine vehicle production feasibility. The professor found that a vehicle could be produced, and at a unit price of less than 150 (1962) dollars. Furthermore, market analysis by the U.N. determined that ten million vehicles could be used initially. The need was clear and the methods workable (Papanek 1984). While feasible and inexpensive from the production standpoint, the project was ultimately abandoned due to environmental considerations.

Another example of production feasibility involves the possible production of two-stroke engines. This is not a new concept. The Saab motor company installed two-stroke engines in its 1960 cars. Today, two-cycle motors can be found in some lawn mowers, outboard boat motors, and chainsaws. The two-stroke engine differs from today's four-stroke engines in that it only takes two piston strokes to compress, burn, and emit the by-products of the fuel. Today's four-stroke engines take two strokes to compress and burn the fuel, then two more strokes to force out the emissions.

According to the Orbital Engine Co., the two-stroke engine would provide 20 to 30 percent better fuel economy, would be 50 to 60 percent lighter, would require 200 fewer parts, and would cost 300 dollars less to produce. Why has this engine not been put into production? The two-stroke engines emit more pollution than the conventional four-stroke engines. While the engine is *production feasible*, current and expected emissions standards prohibit its further development.

However, the recent advent of fuel injection has made the two-stroke engine a prospect for production once again. Ralph Sarich, an Australian inventor, has designed a fuel intake system that atomizes the fuel mixture so it burns more completely. Less pollution is created by the combustion. With further refinements this breakthrough may allow the two-stroke engine to meet emission control standards (Hampton 1988).

Production feasibility is concerned with the capability of producing a product. The need for specialized production methods or equipment may prevent introduction, while technological advancements often enhance the feasibility of producing a product. Obviously, the simpler the production becomes, the more attractive the opportunity.

Stage of Development

Inquire: At what stage in development is the innovation?

Risk is also related to the innovation's stage of development. Mere ideas represent the lowest stage of development, and correspondingly, the highest level of risk to bring the product to the point of actual product commercialization.

The microchip used in computers and other similar devices is contradicting the common fallacy that bigger is better. Microchips have been shrinking in size, while performing increasingly phenomenal functions. The stage of development of certain types of microchips is relatively complete, but, with constant improvements, continually changing.

Engineers and scientists have been able to provide society with better products at lower prices over the past decade. When the cumbersome TI-30 hand calculator was first introduced, it sold for 95 dollars (1970s). Today we can purchase a credit card-sized calculator with more functions for a fraction of the TI-30's cost.

How far has computer chip technology advanced? Engineers at the Oak Ridge National Laboratory have developed an infrared transmitter that can be attached to the thorax of a bee. This silicon electronic chip is approximately the size of a half-carat diamond, using an array of tiny solar cells as its energy source. It emits a beam that can be tracked to assist in determining the migrational habits of bees. The goal of the device is to assist in determining a way to thwart the killer bees coming from Mexico, without the use of pesticides.

But size reduction is not the only advancement in the field of microchip development. NASA's Jet Propulsion Laboratory has developed the Mark III Hypercube, another innovation of supercomputers. This machine's peak speed will be two billion flops (floating point operations).

This level of efficiency is created by the geometric arrangement of the computer nodes. Distances between electrical impulses are much shorter than in a typical computer. Each of the 128 nodes in the Mark III Hypercube will have the power of 25 VAX minicomputers. Accordingly, the amount of computer time necessary for calculations should decrease significantly.

This new computer design could, for instance, influence the time it takes to predict the weather. A meteorologist uses thousands of vector and speed formulas to predict weather patterns. With the increased speed provided by the Hypercube system, forecasts can be developed faster and probably more accurately. This innovation is, presumably, ten times more cost effective than the conventional supercomputer while possessing greater capabilities. But at this time and at this stage in development, it remains very risky. Its *real* potential is yet unknown.

It is important to be cautious with products in early stages of development, because considerable amounts of time and money are usually required to make the product market-ready. Thus, while the field of computer chips is now advanced and the limits of technology are extending, new and risky ventures still exist (Fisher 1988). The key, of course, is to separate winners from losers early enough to focus investments on the former and cut losses on losers.

Investment Costs

Inquire: On the basis of investment estimates and likely sales (level and margin of profit), are investments likely to be recovered?

Investment costs represent the up-front costs necessary to bring an idea to the production and marketing stages. Generally, and unless the business intends to operate at a loss, the investment costs must be exceeded by revenues within a certain time frame. This time period will differ as a function of the financial goals of the entrepreneur. To answer this question, proceed in a step-wise fashion. First, obtain estimates of the total capital likely to be needed to begin production of the product. Likely inputs for these figures would include trade publications, trade associations, and expert sources. Government documents may also prove useful since the Commerce Department indexes capital expenditures and investments as part of their Survey of Manufactures. Total these and other start-up costs. Next, estimate how many units at what price would be needed to recover these investments. Show your assumptions and calculations in writing. If you have an IBM or compatible computer, you may wish to use a software tool, such as Lotus 1-2-3, or Quicken, to help.

Investment costs include all expenditures necessary to begin production and marketing of a product. This is a complex criterion because it can be extremely difficult to obtain accurate information on all possible costs associated with developing a product. Seven common areas of concern are prototype development, pilot plant development (whether to start one's own plant or cooperate with an already existing plant), market research and testing, equipment, personnel, the establishment of distribution channels, and interest (cost of money). In the most usual case, the manager will, unfortunately, underestimate these expenses, requiring additional capital at later stages.

This issue cannot be ignored by the inventor, as these investment costs are directly tied to the profitability of the business venture. The "ingredient method" discussed by Levin (1983) may help entrepreneurs create a reasonable estimate of investment cost. See the following example.

CATEGORY	*EXAMPLES*
prototype development	purchase a competitor's VCR to dismantle and later develop one's own model create a workable car model to test the feasibility of full-scale models
pilot plant development	develop small-scale process for new pharmaceutical (laboratory-level) lease space and equipment off-premises for low volume pro-

	duction of a new soft drink rent a building and develop a workable method to produce childrens' dolls
market research and test	conduct focus group tests on concept of chicken-meat "wiener" carry-out test market (controlled) examining impact of varying advertising levels on demand
equipment	purchase drill presses, band saws, and hydraulic presses to manufacture washers purchase office furniture and a PC to support a consulting firm
personnel	pay employees to run the drill press and create washers hire computer software package specialists to train employees
distribution channels	cost associated with storing steel or rubber washers; sales force costs pay truckers and airlines to transport dolls to toy retailers development of point-of-purchase materials to aid retailers at store level (sale of wieners)

The sum of these and other overhead costs must be obtained, and the cost of money (loans) added. This lump sum over the period of planning (development period) is the investment cost. Managers want to know that the return on their investment will *exceed* what they might achieve through other investment tools (e.g., certificates of deposit, money market funds, or other financial instruments). Thus, investment cost cannot be entirely separated from the issue of payback period or profitability.

Payback Period

Inquire: How long is the payback period likely to be?

This is the proverbial "How long can you tread water?" question of the business world. It asks what the likely payback period will be for the idea. Obviously, shorter is better. Take into account the fixed and overhead costs that will be involved in getting the product to market. As related to investment costs, how long until the venture will break even?

Though robots may be functionally feasible, are they worth the sometimes large initial investment? A robot that can be used in production for welding typically costs 150 to 250 thousand dollars, depending upon the complexity of movement demanded. Will a large investment such as this ever be recovered in savings, i.e., will the payback period never be reached?

The payback period is the time required for the savings to equal the initial investment. The amount of risk involved in the investment can be inferred from the length of this period. A short payback period, say five years, is less risky than a long, (say, 20 years), period. This period also depends on the lifetime of the purchased equipment. If the payback period is greater than its useful life, the investment may not be financially sound.

A useful illustration is seen in robotics. Robots can save untold amounts of money and can improve the environment in which workers operate. The worker benefits from the robot because it eliminates the problems resulting from worker injuries and exposure to toxic substances (e.g., lifting large, heavy production objects; spray painting with lead-based paint).

The company benefits from the accumulated savings. Current health care, retirement, and Social Security costs are reduced. Robots may be more accurate than humans and do not fatigue, which can improve quality and the production scheduling for the product. Simply put, the amount of actual *working* production time will increase. The only care some robots may need is minor daily maintenance and a weekly inspection for component wear.

In cost comparisons, an assembly line employee averages 23 to 24 dollars per hour with benefits. Including maintenance and depreciation, a robot can be operated for 5 to 6 dollars per hour. At this rate of savings it may take about a year of operating time to return savings over a human worker. Robots could be a partial solution to the problem of competitiveness in some American industries. This would be especially true where labor costs comprised a large portion of the cost of goods produced.

Disadvantages of purchasing robots include the possibility that many people may be furloughed due to the increased efficiency of operations. Industrial authors have predicted that by the year 2000, three million such factory jobs will be eliminated in the United States. However, other jobs will be created. Personnel must manually operate some robots, and technicians will be needed to program and maintain this sophisticated equipment. Also, training of existing employees in these fields can offset the large number of jobs predicted to be eliminated (Crocker and Guelker 1988).

Robots are becoming more economical to purchase, since hardware and software costs are declining, thus shortening the payback period. Robot technology has advanced far enough that some companies can recover their costs in a matter of a few years.

Profitability

Inquire: What is the total amount of profits projected for the venture?
Are these profit levels acceptable when compared with the rate of return for other investment alternatives?

Profit is the major objective of almost all American businesses. Profitability of a business venture is affected not only by financial decisions (e.g., the manner in which initial capital funds are obtained), but also by nonfinancial decisions (manufacturing and marketing). Projected over the planning period, profitability addresses the ability to earn money during operation using the acquired capital.

The best way to estimate profitability is by completing a profit and loss statement. This demonstrates whether the venture made might make any money, and pinpoints problem areas. The table below shows a hypothetical profit and loss statement for a consumer product.

Profit and loss statement for a nail polish in its 4th year following introduction (in thousands)

Sales revenue		25,200
Variable costs		9,150
Labor	3,000	
Materials	1,500	
Manufacturing	2,900	
Variable sales commissions	450	
Distribution	1,300	
Assignable nonvariable costs		12,450
Personal selling	600	
Product advertising	8,700	
Sales promotion	2,900	
Salary (product management)	250	
Contribution to profits		3,600
Nonassignable costs		2,350
Corporate advertising	750	
Fixed joint costs		
Manufacturing	900	
General administration	700	
Net profit		1,250
Investment	5,300	
Plant and research equipment	4,200	
Research and development	1,100	

PROFITABILITY MEASURES:

Return on investment (1,250/5,300) = 23.6%
Contribution as a percentage of sales (3,600/25,200) = 14.3%

To review, sales for the product this year were 25.2 million dollars. The variable cost category totaled 9.15 million dollars, with other fixed costs relating to the marketing of the product (assignable nonvariable costs) totaling some 12.45 million dollars. Thus, the difference between sales and the two cost components leaves a difference of 3.6 million dollars, excess contribution to profits.

To completely account for our costs, we must also address the overhead costs of doing business. In this example, these are referred to as the nonassignable costs of doing business. These include such expenses as secretarial assistance, utilities, insurance, and others. They are nonassignable in the sense that they are not directly related to the product's market performance, but they must still be included in the determination of overall corporate profits. Managers usually arrive at a formula for spreading these costs among a company's various products. For example, if two products shared the nonassignable cost figure, essentially cutting it in half, our net profit would go up considerably.

Marketing Research

Inquire: Will market research for this product be simple, or complex?

Every product must undergo market research. Market studies tell us the degree of "fit" between the innovation and its likely markets, level of customer need, and the strengths/weaknesses of the competition and their product offerings. In general, at the screening stage, secondary research is used to identify the likely markets, profile the buyers/end-users, and get a feel for the competition. At the next level, many consumer concepts can be tested relatively cheaply on a very preliminary basis with small samples, frequently costing less than 10,000 dollars. A complete market test (including introduction under controlled conditions) can run as much as several million dollars. In order to answer this question, consider the degree of complexity of the technology being discussed, the familiarity of the target market with the idea, and the extent to which the innovation has been developed.

Before jumping into a swimming pool, most people test the water by inserting a toe. This helps avoid the sudden shock of unknowingly diving into what might be icy water. Similarly, the entrepreneur should test the market before swinging into full production. Market research is utilized in determining the market potential of a product, and in following-up with customers.

Duncan, a parking meter manufacturer, spent months determining where its meters could be used in Europe. The first step was to find places that would benefit the most from Duncan's product. Congested streets were located by using secondary information (city records) to identify areas that recorded increased populations and vehicle registrations. Parking regulations where congestion occurred were studied, and photographs were taken to present to authorities to show how cities could benefit from using the meters.

The meters eliminate disorganized parking, while also generating substantial revenue. Quality and variety also contribute to the marketability of the meters. Duncan meters are specifically designed to prevent their theft or vandalism. Eleven models can accommodate 15,000 time and rate combinations. Any size coin can be used, and instructions for meter operation can be printed in any language. This ability to customize the meter to individual countries' specifications significantly expands the meters' potential market (Rosenthal 1988).

Likewise, Kleenex could not rely on name alone to sell its product; it had to *develop* its market abroad. To sell its facial tissues in Japan, arrangements had to be made for Japanese companies to manufacture them. And, differences in consumers' needs for the facial tissue needed to be examined. In America, Kleenex tissues are used primarily as nose wipes. The Japanese, however, find such direct promotional appeals offensive. The product had to be positioned for use as cosmetic tissues, not nose wipes. The box design was also modified to appeal to the Japanese consumer.

Following these tasks, product distribution became the primary concern. Though a devalued dollar contributes to consumer appeal (U.S. products are less expensive overseas under these circumstances), the company recognized the need to establish relationships with its overseas manufacturers. Price alone would not be enough. American executives developed expertise in the new export-oriented environment, physically traveling to and studying the Japanese market. Kleenex could not have been effectively sold from the states (*Marketing News* February, 1988).

Developing a market overseas is time-consuming and expensive, but the possible rewards can make it a worthwhile pursuit. Market research is a necessary step in both domestic and overseas settings.

Research and Development

Inquire: Will the research and development period be long and difficult, or will it be short and straightforward, relying on known technology and expertise?

The level of rescach and development required is usually an important factor in bringing an idea to commercialization. The lengthier the development period, the longer capital and resources are not productive. More difficult concepts can take longer to bring to market and thus involve more risk. Generally, companies are willing to make trade-offs: long research and investment periods must carry bigger payoffs or firms will quickly shelve the idea.

American consumers change their tastes and preferences so rapidly that it may be hard for a company to maintain an acceptable overall market share. Technology must accelerate beyond the pace of the consumers' tastes and preferences, *anticipating* in order to satisfy consumer wants and needs. Research and development departments must continually innovate to meet the demand. Consequently, new products are continuously being introduced into the market. During 1988 alone, more than 5,600 new consumer products were launched, along with some 10,500 new variations of existing products. And, according to Marketing Intelligence Service, Ltd., a record 15,866 health, beauty, household, food, and pet products were introduced to U.S. consumers in 1992 (Miller 1993).

In recent times, over 25 percent of 3M's sales have originated from new products. The innovation race is an obsession which can only be satiated by research and development, but it also carries a large entry fee. The willingness of U.S. companies to invest vast sums of money into research and development is one of the reasons the country has been a world leader. Since this lead has been slipping relative to some other nations, it has been the subject of considerable concern.

The key to focusing research projects is to develop commercial instincts in company scientists, making them consumer-oriented. Though new inventions are an asset, they are not practical until consumers have a use for them. An optimal research and development employee would be a scientist who could not only discover a new product, but who could identify practical uses for it in the marketplace.

The link between new ideas and their markets is often not clear in early stages. For instance, SmithKline Beecham did not want to introduce a hepatitis vaccine into the U.S. market, because litigation involving unreliable pharmaceutical drugs was so common and costly. The company's researchers, in turn, took the product to the firm's international distributor. It was introduced in limited international markets, and was an immediate success, resulting in sales of 50 million dollars for the drug in 1980 alone (Swasy 1989).

In another instance it was marketers who informed the research and development department of a product need. Corning was trying to sell safety windshields to General Motors. Although the windows were not sold, Corning marketers were informed of an urgent need for an anti-pollution device. This need was relayed to Corning's research and development department.

There, scientists had created a porous ceramic material the previous year, but no use for the product was foreseen at that time. However, the engineers were able to utilize the ceramic material in the development of a catalytic converter, form it as GM designers specified, and meet environmental standards. The device satisfied the need for a catalytic converter in the auto industry, resulting in 150 million dollars in annual revenue for Corning during the late 1980s.

Innovations may be discovered haphazardly, but such accidental discoveries are especially valuable when consumer benefits are clear. For example, a Raytheon engineer was working on experimental radar equipment when he noticed that a candy bar in his shirt pocket had melted. Further investigation of this occurrence led to the development of the microwave. A researcher at 3M spilled an industrial compound on her shoe, and after some time noticed that her shoe was clean where the compound had splashed. With improvements came the highly successful Scotchgard brand fabric protector (Labich 1988).

Organized research and development has been important to discovery and development of new products. Managers also realize that R&D is necessary and expensive, and therefore remain skeptical about new products until their potential to satisfy some need has been demonstrated.

Demand Factors

Demand for a product is the demonstrated level of interest expressed by the marketplace for that product. In this section, attention is focused on those aspects of demand that should be identified very early in order to determine if sufficient market opportunity exists for the product.

Market Potential

Inquire: What is the time period for market assessment?
What are the product's characteristics, and how well will they endure in the marketplace?
Will the product be positioned as a new brand or new product?

Assessing market potential is neither easy nor a certain process. It relates to the maximum possible sales opportunity available for all sellers of a good or service. This is, of course, constrained by the level of competition, the impact of new or changing technology, the time frame being analyzed, and the nature of marketing strategy used by the firm(s). In the latter case, a manufacturer might decide from the very beginning not to focus on the mass market, but to concentrate on a small niche of buyers. In this instance, the near-term market potential would be considerably smaller than if the firm was willing to assume greater risks and attempt to capture all possible sales.

For several decades IBM monopolized the electric typewriter market. The market for their "Selectric's" has deteriorated considerably, however, owing to the inroads made by word processing equipment. Thus, as a result of the greater flexibility and parity of costs associated with producing and using the newer technology, the market potential for IBM's "Selectric's" has diminished. The manufacturer of any electric typewriter now must compete very strongly on performance and cost dimensions to be able to enter this aging typewriter market effectively.

The market potential for CD Rom computer disks is improving daily. Instant access to computer files of very large size and complexity is appealing to research oriented users. In addition, CD Rom disks offer audio (music) and video selections that can be played during computer use. The standard 5 1/4 inch computer floppy disks are themselves now becoming obsolete. Since the introduction of the more durable and smaller 3 1/2 inch floppy disks, and with wider access to cheap hard disk installation, the market for the older disks is diminishing quickly.

Determining a product's market potential involves carefully identifying every conceivable substitute in the market sector. A "universal" product will have enormous market size. Conversely, a product designed for an exclusive market segment will have a limited potential market. For instance, a glass bottle manufacturer's only competition is not other bottle manufacturing companies: plastic, tin, and aluminum producers all make comparable substitute containers. All competitors must be considered when analyzing the glass bottle's market potential.

Minoxidil, a drug that stimulates hair growth, is directed at an exclusive market segment. Approximately 30 million people, primarily male, are bald or balding. Though the market segment is restricted to bald people, the market potential is high because of the uniqueness of the product and the initially large number of potential users (Smith 1986).

Prior to this drug, the solutions to baldness were hair implants, hair pieces, or hats. "It's (minoxidil) the first product that actually promotes substantial regrowth of hair in bald areas," reports Dr. Thomas Nigra, chairman of the dermatology department at Washington Hospital. Rogaine, the trade name given to the minoxidil cream, was investigated by the FDA for possible side effects. The product was approved in 1990. Upjohn, the drug's manufacturer, has released Rogaine to the public and has already experienced moderate success.

Potential Sales

Inquire: How many likely buyers?
What share would the product garner, and what would be the rates of retention and repurchase?
What's the planning period?

This question relates to the size of the market for the innovation. You should consider, therefore, whether competitors have shares of the same market, and what those shares are (or are likely to become), and then estimate the volume of units likely to be sold in the planning period.

All of this is easier said than done and involves considerable "guesstimation" and research. In fact, this part of the assessment is very dependent upon identifying the segments of buyers, counting the numbers of buyers in those segments who would buy initially, and then estimating the rate of repurchase over the planning period. This task will be the same for industrial sellers as for marketers of consumer products. Thus, some market research on the behavior and profile(s) of buyers is necessary.

Suppose that market research for double-sided, double density diskettes shows that new purchasers in one segment, the educational setting, typically buy 500 per year. If there are about 2,000 such institutions buying the diskettes, the total annual initial market is about 1 million units. If research suggests that the new brand will be able to garner about 10 percent of this market, the initial sales potential will be about 100,000 units. If we summed the total across all segments that would be reached (office supply, large industrial buyers, government outlets, etc.), we would be able to total the sales potential (having taken into account other competitive shares and entry level marketing effort assumptions). If we review second year potential, we would have to take into account such things as brand loyalty (expressed as the percent of sales that we would be able to expect from satisfied customers), and the emergence or attrition of market players. Hot markets frequently attract new competitors in the second and third years, and may subsequently experience a decline in number of players, depending on the condition of the market.

The producer of a radically new product for which there is no substitute or brand competition is faced with a much different task. The analyst must determine the rate at which the product would diffuse during the first year, making some assumptions about this depending upon market research. Again, careful definition of target segments and the intensity of need for the product will enter into the equation. Consider the task faced by manufacturers of a product designed to extend the life of contact lenses by 50 percent. The user-market is large and likely to be receptive to the idea, but there will be some resistance among distributors, where the innovation must first be accepted. The first market would be the "trade" (optometrists and opthalmologists), since they are the only initial way to reach the end-user, in this case personal users.

Strategically, large consumer-goods marketers often initially examine the mass market (very large number of users), and then smaller niche markets. Small companies may be wiser to think niche first, then, as resources permit, mass market.

The push to ban cigarette smokers from public places has increased demand for products designed to limit the urge to smoke. Where hypnosis, filters, and shock therapy have failed, nicotine patches have grown in popularity.

Nicotine patches deliver decreasing levels of nicotine through the wearers skin until physical dependence on nicotine all but disappears. The adhesive patch is relatively convenient in comparison to other methods, and can be easily hidden beneath clothing. The patch has captured a large portion of the market segment interested in kicking the habit. The potential market for nicotine patches was predictably large because of the current number of tobacco smokers in the U.S. (approximately one-quarter of all adults).

There are four main brands of nicotine patches, Marion Merrell Dow's Nicoderm, Ciba-Geigy's Habitrol, American Cynamid's Pro-Step and Warner-Lambert's Nicotrol. Expected sales levels are reaching into the millions for 1992. Increasing pressure to ban smoking in all public places will cause the potential sales levels for smoking control devices to increase correspondingly. The four giants in this market are expected to increase their combined advertising campaign to the $100 million level in 1993 in an effort to improve potential sales for nicotine patches (*Advertising Age* December, 1992).

Trend of Demand

Inquire: Is the primary demand curve for the product category rising or falling?

This criterion relates to the prevailing market trend, since few markets can be classified as strictly stable. In other words, the market demand for a product is frequently on the increase (growing by x percent) or on the decline (dropping by x percent). For example, in the case of typewriters (discussed earlier), the introduction of word processing software, targeted to the small business, contributed to the decline in demand for older mechanical typing products and technology. Additional examples of products in decline include fish ties, saddle shoes, and LED watches, while products in an upward swing include LCD watches, business accounting software, and computer games.

If the sales levels for confectionary items were plotted over the course of the year, the graph would reveal that dramatic increases in sales are associated with holidays. Until recently, even major candy manufacturers largely ignored this seasonal trend. Today Hershey Chocolate USA, Mars and Nestlé Chocolate & Confection Company try to prepare new product releases and special packaging for the holiday season to increase their products' visibility.

By concentrating promotional efforts nearer to holidays, sales have been steadily increasing for the past few years. Christmas was found to be the second-best selling holiday, with 8.6 percent of all candy sales, while Halloween takes top billing for the fastest peak in candy sales. Confectionary manufacturers expected Christmas sales in 1992 to increase 2 percent over 1991 sales by 1.2 billion dollars because of strategic new product releases and holiday related packaging. Total confectionary sales are expected to increase 4 percent in 1992 to 14 billion dollars, according to the National Confectioners Association (C. Fisher 1992). Note that this growth is taking place despite Americans' increasing concerns about diet and health. Clearly, the trend in demand is favorable for this category of goods.

This may not be the case for manufacturers of some distilled spirits. During the 1980s, demand for many brown spirits (e.g., whiskies) declined. Consumers were switching to lighter, lower alcohol beverages such as gin and rum. While there is some evidence that components used in cocktails go through 15-year cycles, some producers fear that changes in lifestyles and attitudes toward health and diet may mean continued declines or, at best, a flat demand line for many distilled spirit varieties. Thus, the trend in demand for some is not expected to be favorable (Winters 1986).

Stability of Demand

Inquire: What factors affect the demand for this product?
Are any of these factors controllable or predictable when estimating how stable the demand for the product will be?

Simply put, this factor deals with whether the demand curve is stable or fluctuates as a function of factors that are known and helpful to prediction. Beer industry sales, as well as soft drink sales, are stable in the sense that they are seasonably predictable: higher in the warm months, lower in the cooler months. This is unrelated to the fact that beer sales are essentially flat, i.e., their rate of growth or decline is very gradual. The current trend for beer consumption is in fact, decreasing. In 1991 the beer industry experienced a 4.3 million barrel (2.2 percent) decline in sales (*Marketing News* February, 1992). Thus, while the trend may be a declining one, the overall pattern of sales is quite stable, indeed, highly predictable. Demand is strongest in the warmer months, weakest in the colder months, and can be held constant by such factors as holiday periods, region of the country, and/or major sporting events.

On the other hand, demand for certain medical isotopes is very unstable. It is virtually anyone's guess just when a new compound will be discovered that will replace an existing one for diagnostic tests. New, high-technology equipment is developing at a breath-taking pace; what was new will soon be outdated. Highly unpredictable products may include isotopes, strategic metals, or custom defense products; moderately unpredictable cases would be gasoline markets, raw goods such as flour, concrete, and steel; relatively predictable cases would be shoe markets, clothing (general, not fashion), milk, and soft drinks.

Snowblowers are a popular product in northern states after the first snow. Manufacturers are producing numerous models in a wide range of prices. Retailers can afford to stock snowblowers because the demand is fairly stable (it usually snows at least once in the North each winter).

Retailers realize that snowblower sales are driven by impulse and desperation. The best deals occur either before the first forecasted snow, or when the lawn-mowing season begins. Thus, snowblowers are seasonal items with reasonably predictable demand. Certainly, less risk is involved in the marketing of these products. The amount of snow received is, unfortunately, not always so predictable (*Flower and Garden* 1987).

Demand in the clothing industry is less stable. Fashions change every year, sometimes every month. For example, color trends over a five year period have progressed from black to green to brown and back to black. Styles have varied from trim to loose-fitting to polka dots to clashing. A retailer who purchases an abundance of a particular fashion, but is unable to sell the stock while it is popular, may be forced to sell it below cost. The more trendy the item, the higher the risk, but the higher the risk the higher the retail price.

Thus, products exhibiting sales which are highly unstable and which fluctuate unpredictably usually indicate higher levels of risk. If retailers are not convinced that they will have a high rate of merchandise turnover, they will not be willing to buy large quantities of the product.

Length of Product Life Cycle

Inquire: At what stage in the product life cycle is this product? What risks and tasks are associated with this stage?

There is some evidence to suggest that products go through somewhat predictable life cycles: an introductory stage, a growth stage, a maturity stage, and a decline stage. Some products, because of the degree of risk, amount of capital, barriers to entry, underdevelopment of competitive technology, etc., seem to go through very long life cycles, i.e., the products and demand levels remain relatively unchanged over long periods of time. Examples of this would include machine tools (the tools used to make many home appliances), facial tissue, basic socks, etc. None of these items is subject to great innovation or rapid change in the technology. Likewise, each serves a very basic need, not likely to change dramatically in the long run.

Manufacturers prefer that the life cycle be longer rather than shorter so that they can take advantage of the learning curve–their own experience in cutting costs and reaching markets–to maximize market shares and profitability. Products with long life cycles include clock radios, baseball gloves, ball-point pens, and lead pencils. Items such as "power" suspenders, and high fashion items like eyewear usually have a shorter life cycle. Note that in the case of eyewear, no product has completely replaced basic eyeglasses. The style may change, but the refraction basis remains. Such fad products as Treasure Trolls™ are inherently more risky ventures. One must get in and out quickly before the fad collapses and the manufacturer/distributor is stuck with inventory and losses.

The time span of each stage of the product's life differentiates each product's marketing needs. If a short life span is projected, intense promotion and visibility is needed. This is especially evident in "trendy" items.

One product which may have a lengthy life cycle is the battery cell. Competition in this market exists, but the battery life cycle is far from decline. The basic product is constantly undergoing development, but the basis for demand (provision of portable power) is a very strong motivation for continued purchase. The market is profitable for those companies already in competition, and promotion is a major tool.

Kodak's Gold Tip is supposed to outlast Duracell by at least an hour, while Duracell claims a mechanical bird chirps 20 percent longer with a Duracell battery than with a Kodak battery. "No battery in the world outlasts" the Everready Energizer, it is asserted (*Consumer Reports* November, 1987), a refrain nearly all can recite from memory.

The market offers millions of products that need batteries to operate. The Sony Walkman requires batteries. Remote controls, "boom" boxes, hearing aids, garage door openers, and watches all need batteries to function. Yet with so many new products, markets for new battery products are constantly appearing. The product fulfills a very basic need in contemporary society.

Product Line Potential

Inquire: Are there related applications in the same markets?
Are there different markets that may be interested in this product if slightly repositioned or reformulated?

Is this the only application of the product? Or does it have others (perhaps with slight modifications) in other markets? Are there some related products that could be developed that have some potential to introduce marketing economies of scale (e.g., similar buyers) or that would utilize existing technology in new ways? High product line potential prospects might include movies ("Beauty and the Beast" or "Aladdin" characters and toys), WWF wrestling characters (the "Ultimate Warrior"), baking goods (several baked goods from the same ingredients), microcomputers (new and faster models incorporating peripherals such as FAX cards and CD readers).

Product line potential is the value of the product in terms of possible spin-off products or multiple uses for the product in different industries. A product that has had success in creating spin-off products is adhesive tape. Today we have masking tape, duct tape, clear tape, transparent tape, electrician's tape, and so on. Another product that is experiencing success in numerous industries is barcode readers.

Barcoding technology over the past ten years has penetrated almost every conceivable phase of the marketing process. Barcoders originated in the manufacturing industry as a means of keeping track of inventory. The identification bars were stamped on the item itself, or placed on a tag and fastened to the container. The scanner passed over the identification bars, relaying the information to a data base where the inventory information was stored.

The usefulness of the infrared scanner was recognized almost immediately by other industries. Product line potential for uses other than inventory control was investigated by innovative marketers and engineers. Other industries began to conceive different uses for this technology, measured in terms of new applications in existing markets, as well as new markets for existing applications.

Today, barcoders can be found in the consumer marketplace as well as the manufacturing industry. They have virtually invaded large retail stores, grocery stores, and libraries. Most large retail and grocery stores use barcoders to enhance the speed of the purchasing process. Instead of "punching in" the price of each sale, the clerk drags the item across a scanner that identifies and prices it. Libraries and video stores use the barcoder as a more efficient way of tracking their items (Schwedelson 1988).

The concept is spreading to other industries. Caterpillar has developed barcode transportation vehicles. The vehicles are guided by barcode programs, which are read into the buffer memory of the vehicle from a routing chart. The vehicle's routing memory is stimulated by different barcoding patterns.

Instead of finding new applications for a product, a different strategy would be to differentiate the product. Most diapers perform the same function, but by marketing a revised product in different ways (repositioning it) across different market segments, the product line can be expanded (i.e., infants, toddlers, adults). Kimberly-Clark Corporation introduced a diaper targeting small children who may have "outgrown" Huggies^{TM}, but who are clearly not ready to go solo. During the summer of 1992 Kimberly-Clark introduced Pull-Ups; training pants for children three and over. Since the introduction of Pull-Ups, other companies are planning to launch their own line of small children's diapers: Procter & Gamble's (First-Pants) and Weyerhaeuser Company's (Undeez, Kinderpants, and Wonderpants).

Drypers Corporation is expected to further segment the diaper market by introducing gender specific colored diapers. By expanding the diaper products offered, diaper manufacturers can control the shelf space in retail outlets making it too difficult and expensive for new competitors to enter the market (Lawrence 1992). Thus, the longer term strategy of all these firms is to place a lock on the range of segments that have needs for these products, fully satisfying the needs of all potential users.

A product with a high product line potential is desirable. If one product begins to lose market share, complementary products can maintain profits and share overhead costs. The inventor should be aware of the options available for his or her product. With potential spin-off uses or products, a product becomes more appealing to a manufacturer. At the pinnacle of this principle is the Campbell's Soup Company. Is there a soup they *have not* considered?

Market Acceptance Factors

Many products are introduced each year, some literally "take-off" like rockets, rapidly gaining widespread acceptance, while others never seem to catch on. In this section attention focuses on those factors that affect the rate of acceptance of new products, for better or worse.

Compatibility

Inquire: How similar in use, appearance, and function is the innovation to the methods or product in use?

Are existing attitudes/values consistent with those proposed by the adoption of the innovation?

Consumer acceptance of a new product is likely to be greater if it is similar to what it replaces in appearance or use. The lesser the adaptation needed, the easier the transition will be. Under most circumstances, this improves the likelihood and degree of acceptance.

Convincing consumers that they should purchase a heating system that is built on the principle of hot-water panels may be difficult. Most heating systems are built around the notion of moving hot air through ducts. In homes where steam heat was the norm (radiators), the task would be easier, as the buyer's perceptions of how to heat the home would already be more compatible.

Consumers have a preconceived use for most products on the market. If consumers are not familiar with a product that a manufacturer introduces, they may not be able to envision its uses. The product must be compatible with consumer usage patterns. If not, heavy educational costs might be incurred in advertising and promotion.

The first computers introduced to the marketplace were specifically geared toward the technical professions. Chemists, mathematicians, and engineers used the computer for complicated calculations. The need for a computer in a variety of industries was not fully realized. Making the computer compatible (user-friendly) for society in general was not even considered.

Today, computers have expanded into all realms of society, not just large companies able to afford mainframes. The personal computer (PC) has contributed greatly to this expansion. Since the price of the PC is a fraction of the cost of a mainframe, smaller companies can afford to use the PC in many of their processing procedures. Similarly, multiple 486DX computer models can offer the same speed and memory capacity as some supercomputers of the 1980s, with significantly more flexibility, greater reliability, and less technical attention.

Many people are eager to learn how to use the PC, but they do not want the traditional training involving hardware and software. Most often, people want to learn how to use the PC for a specific task. In the word-processing application, for example, it was critical that the keys on the computer keyboard mirrored as closely as possible the traditional typewriter key pattern. Without this feature, rate of acceptance of PC systems would be a fraction of what it is today.

Nontraditional methods have been developed to cater to this growing service industry market. Training is specifically geared toward the trainee's job. This typically involves a "hands on" introduction to the hardware and operating system, and then focuses on developing skills useful to the tasks required on the job (Minicucci 1987).

Learning

Inquire: Does it appear that this product/service will be easy or difficult to learn to use?
What is the requisite level of knowledge/experience to use this innovation correctly?

Initially microcomputer companies had to overcome the appearance (one for which they had been smug for years) of technical complexity. This image was somewhat acceptable when the primary markets for the product were hackers or "techies," but when trying to reach the larger portion of the market with *no* experience this was a liability. Simply put, processes or products that are easy to use or easy to learn are likely to be accepted more rapidly than those that are relatively more complex or those that may require extensive training. Products difficult to learn to use include microcomputers and highly technical software. Easier products are exemplified by electric coffee makers, live-capture animal traps, televisions, and fishing poles.

A 35-mm camera once was the trademark of only the professional photographer, because the average person had difficulty comprehending aperture, F-stop, film speed, and light meter settings. Learning all of the concepts necessary to properly use standard 35-mm equipment was a major barrier to mass marketing the product.

This difficulty influences a consumer's decision. Innovations that reduce learning requirements will make a product more appealing.

Kodak and Minolta are leading innovators in the camera industry simplification process. How simple? An image motion detector automatically sets the focal distance, a task that used to be done by lining up an optic circle in the viewfinder. Usually, if the exposure is too dim or bright, a flashing light is triggered in the viewfinder, alerting the user to change the setting. The film speed is set when one loads the film in the back of the camera. An amateur can thus take high-quality pictures with relatively little acquired skill. The user has only to load the film, point the camera, and press the shutter button (Goldberg 1981). It has become as simple as point and press.

Need

Inquire: Can segments of consumers, that may need a product, be clearly identified?
Does the product fulfill essential needs, or are these frivolous?
Is the need physical (more basic) or induced psychologically (must be instilled by promotion)?

Some products come and go rather quickly, not because they are not easy to use, but because there is no real need for them. Pet Rocks came and went in a matter of months. One could argue that they fulfilled no real need for the buyer. On the other hand, Liquid Paper brand cover-up for typing errors has been on the market for decades, as real need exists among users lacking good typing skills. Need is relative, in the sense that different users express different levels of need for the same products. When comparing large groups of products, consumers would probably agree that they need some more than others. One may have little need for pet rocks, pet spiders, most shelf knick-knacks, paperweights, and soft drinks, but have a strong need for automobiles, detergents, shoes, and eyeglasses.

The glass manufacturing industry has recently been overrun with automobile manufacturer requests regarding design and composition of windshields. New design requirements in the automobile industry are forcing competitive glass producers to react quickly to fulfill shifting customer demands. Feasible processes had to be developed to form glass into various shapes while still providing strength.

Larger glass surfaces are needed for the new car body designs. Though this may appeal aesthetically to the customer, it can pose some challenging consequences for the safety and comfort of the passengers. Glass development is also affected by advances in the electronics of the car. For instance, when electric defoggers are installed within the rear glass, extra stresses are imposed on the bonding characteristics of the glass. And, if the rear glass is contoured, further complications arise.

Glass is a heavy material. Increasing use of glass in the design structure adds weight and cost to the car. Maximum strength and ductility with minimal weight is desired. A tempering technique has, therefore, been developed that strengthens the glass, satisfying the need for the glass to be thinner but as strong or even stronger than the previous type of glass.

Because glass is transparent, solar energy passes through it producing excessive amounts of heat. The simple solution of past years was to tint the glass, thereby deferring the penetration of rays, but car manufacturers needed a more efficient solution. Electroconductive films of noble metals, that can reflect solar radiation before it penetrates the car, have been developed, and are applied like a film to the surface of the windshield.

When struck by sufficient force, glass shatters. Even though a windshield might prevent ejection of a passenger, the person may experience disfiguring facial lacerations. An antilacerative windshield was developed to alleviate this, using a thin polymer film bonded to the inner surface of the windshield to contain glass fragments.

Glass shaping has become complex. Improved aerodynamics drive the need for techniques that bend glass as well as apply it to the car body. Special molds create the necessary bends while preventing optical distortion. A polyurethane adhesive is then used to attach the glass directly to the car without numerous moldings (Colombin and Jourdaine 1988).

It is evident that the needs of the consumer can stimulate new, innovative creations. The entrepreneur must be in touch with his or her market to understand its needs. Careful study will usually reveal the level of need for a product. The inventor's job is to apply available technology to formulate a solution to satisfy this need.

The entrepreneur must, therefore, recognize that manufacturers are more interested in products that fulfill consumers' needs than in fad products that involve high investment risks. Producers must be consumer oriented.

Dependence

Inquire: To what extent does this sale and subsequent use of the product depend upon ownership or purchase of other allied products/services?

This criterion taps the degree to which the sale depends upon the user's possession of other goods or services. For example, when IBM originated the card keypunch machine, it also *required* its buyers to enter into tied contracts. These contracts guaranteed that the buyer of the keypunch technology would purchase only IBM brand keypunch cards. The contracts were subsequently outlawed as unfair, but most products can still be tied in some practical fashion to others.

The degree of dependence of one product on another influences market control and penetration. For instance, the sale of record albums is completely dependent on the number of people who own a record player. Comparably, the only medium that will work in a compact disc player is a compact disc. And, persons who purchased Sony beta VCR equipment know first-hand of the interdependence of the goods. This in fact, slowed rates of product adoption.

Solar powered products are dependent on photon energy. Solar panels convert light energy into electrical energy by using the photons' kinetic energy to transfer electrons through a series of outer valence shells. Although these solar products never require replacement of a battery, they are dependent on a significant amount of light (photon energy) to operate, without which the product becomes virtually useless.

Market penetration levels may be limited or slowed if another product or service is required for use of the proposed product. If a product is virtually independent, the risk involved with future use is of a different type, involving systematic efforts to respond to competitors. A compelling example exists in the difference between Apple and IBM microcomputers. The two systems have been constructed around different hardware and software architectures, and it has been only recently that communicating between the two has been simplified. Certainly, this prompted some organizations to be sole users of one or the other, with little room for compromise.

Visibility

Inquire: Will the product be a visible reminder of need to use/consume?
Are the benefits of using the product visible/perceptible to the consumer?

Generally, products that are more visible in action are adopted at a higher rate. If, after using D-CON brand rat killer, a farmer sees that rats are dying, then visible proof exists that the investment was worthwhile. Highly visible products include dishwater detergent, toothpaste for smokers, auto wax, vacuum cleaners, and the umbrella. If performance is visible, it tends to serve as its own reminder of the need to use/consume. This helps speed adoptions and results in satisfaction and repurchase. Visibility represents the marketer's ability to communicate the advantages of a product to the desired target market.

Skin cancer is among the latest health concerns of the consumer market, with incidence rising nationally. People have turned from "sun worship" to "sun fear." No one knows this better than the suntan product manufacturers. As people became more aware of the negative health consequences of tanning, the market shifted from suntanning enhancement products to suntan prevention products.

Sales of most protective products have thrived since the public has become aware of solar dangers. Physicians and the press provide constant encouragement to society to shield themselves from the sun. Consumers are buying the sun protection products to prevent skin cancer, wrinkles, and critical burns, and, at least in the short term, results are obvious.

Manufacturers continue to innovate successful sunscreens. Now sunscreen systems include moisturizers and enhancers. Specialized products for different body parts (e.g., Noskote for the nose) are also expanding the market. Differentiation for age categories, colors, and fragrances have added to the various strategies to induce consumers to buy more (*Consumer Reports* June, 1988).

Promotion

Inquire: With stated sales objectives, what will be the likely amount of promotion money needed to achieve those goals?
Is the sale generic or competitive, i.e., am I selling the product or the competitive benefits of my brand?
What types of promotion will be needed?

Promotion is the stimulater, the accelerator of sales for the product. It can come in many different forms, ranging from advertising and public relations to coupons, samples, and sweepstakes, to name but a few. Generally, a plan is developed that outlines the nature of the promotion to be used at different times in the product or brand life cycle. This usually takes into account the segment(s) targeted, level of accessibility, the nature of the product, etc. The dollar amounts budgeted depend upon the overall objectives the manager has in mind.

Since many products are not sold on a face-to-face basis, it is frequently the promotional effort that is used to position the product or brand effectively and competitively in the mind of the buyer. This can be expensive or inexpensive depending upon the size of the market, the number of competitors, their current expenditures, and/or the nature of demand. Products requiring heavy promotion relative to sales include most consumer goods (mass market), such as beer, cologne, and soft drinks. Light promotion relative to sales is the rule for durable goods destined for the mass market (cars, homes, refrigerators). In the primary demand situation (no competitors, establishing demand for the product category), consumers must be educated, which can be expensive. In a competitive demand environment, the focus is on brand benefits, which may be less costly to communicate.

Promotion is necessary to inform market segments of new or innovative products. Without paid promotions the firm will have to rely on publicity, essentially uncontrollable by the company. The individual or business must decide which promotional mix will best suit their product's need. When advertising, it is crucial to use the most effective message and cost-effective media to reach the desired market.

The most economical medium, and sometimes the best, is word of mouth. Customers telling their friends about the superiority of a certain product can positively influence additional customer inquiries and sales. The converse is also true, however.

Newspaper ads or yellow pages (directory) advertisements can increase the portion of the market reached. For a few cents per word an entrepreneur can reach as many customers as there are subscribers. Small, local businesses usually choose one or both of these options, not having the cash or sometimes even the desire to reach a larger market. Nationally published magazines or major commercial network television advertisements can reach the entire universe of adults, and major consumer goods firms spend millions of dollars to promote their merchandise through these media. The largest spend in excess of 1 billion dollars annually.

Advertising is big business in the United States. As international competition for market share has intensified, major manufacturers concentrate on product or service differentiation through advertising. The first battle is for the consumer mind, with advertising being critical in establishing a favorable product or service image. Despite the downturn in automotive sales in 1991, the automotive industry spent 5.259 billion dollars on advertising. Likewise, retail stores spent over 5.1 billion dollars in the effort (*Advertising Age* September 23, 1992).

Individually, Wal-Mart spent 189.8 million dollars in 1991, according to Crain's *Advertising Age* to continue its commitment to low prices and its "Made in the U.S.A." image. This "rolling back our sleeves" image has helped sustain Wal-Mart's growth over the past decade.

Domino's Pizza realized that its sales were slumping since the middle of 1990. In a growing home delivery market segment, Domino's market share was being affected by Little Caesar's "Pizza-Pizza" and intense price competition in the market. Domino's had suffered since their trade character "Noid" was dropped.

From 1988 to 1992, the company has run advertising campaigns without emphasis on differentiation. Domino's promoted fast, hot delivery but did not mention its product or product quality. In the fall of 1992, Domino's tried a more aggressive advertising campaign by tying-in with Nintendo of America and capitalizing on the sales promotion value of popular Nintendo products. This distribution of Nintendo game cards led to the best sales weeks of the year.

Domino's plans to spend its advertising money differently in 1993. The company is concentrating on print budgets and prime-time television, rather than late afternoon. "We're totally changing the company with an expanded menu and an aggressive, recognizable [marketing] direction," states Mr. Sheehan, Domino's president. "We're putting a face on Domino's" (Cortez 1993).

In addition to traditional print and television media advertising, many car manufacturers have used sponsorship in their advertising campaigns. Car manufacturers feel the expense of being an "official" sponsor, is worth getting their vehicles in front of the right kind of people in the right kind of setting.

According to *Ward's Auto World* (Gray 1986), pace car sponsorships tend to help. Oldsmobile has been a frequent pace car at the Indy 500. Robert Jarboe, director of marketing at GM says, "People go there to see autos perform and it's good exposure to have your product out in front of 33 racers."

Subaru has been a sponsor of the U.S. Ski team since 1976. Subaru claims that the popularity of their 4-wheel-drive vehicles among skiers is due to this sponsorship. For this product, this tie-in seems to make good sense. Similarly, Toyota has linked its trucks with Opryland in Nashville, because it views country-western fans as truck enthusiasts (Gray 1986).

The details involved in promotion must take into account the nature of the product and the audience to be reached. Advertising agencies thrive on solving these problems and can help from problem definition, through to the point of evaluating the effects of the campaign.

Distribution

Inquire: Are distributors for the product readily available and established?

Will the costs associated with distribution be reasonable,

or will these expenses drive the price of the product out of reach of the target market?

Channels are used to get the "goods" to consumers, sometimes called distributors, wholesalers, or jobbers. These are the individuals and companies between the manufacturer and the buyer. Good products at the "right" prices are common; yet, if the manufacturer has no formal system of delivering the product to the consumer, the effort may still fail. Consider the cost and difficulty of establishing distributor channels when addressing this question. Where the market exists (geographically), and what customers perceive as the "usual" distributors are major determinants of the system of distribution.

The innovator of a new medical imaging device (electronic scanner) may have a difficult time distributing it. Persons/companies skilled in the technology with connections to major institutional buyers may be difficult to find. The product may have special transportation requirements, and if this is expensive, it might be difficult to persuade distributors to maintain any inventory.

Alternatively, the innovator of a new fishing line may have less difficulty with distribution. Numerous outlets and independent jobbers exist in this industry, and since cost is low, maintaining inventory is not likely to be an issue. Rate of inventory turnover will also be favorable.

The channel of distribution, the route and sequence a product follows from producer to consumer, depends on the product type. Authors Patrick Murphy and Ben Enis divide approaches to distribution into four categories based on the type of product. Convenience products (food, industrial supplies), preference products (chewing gum, auto parts), shopping products (women's suits, cars), and specialty products (industrial machinery, Rolls Royce automobiles) differ, requiring alternative distribution systems (Murphy and Enis 1985).

As one progresses from common products to specialty products, the number of distribution outlets decreases. A niche market may not imply sufficient sales to encourage numerous outlets. Comparably, convenience and preference products should utilize an intensive distribution system, involving many intermediaries and retailers. Shopping products should use a selective distribution system with only a few distributors, since the level of demand will not justify very many outlets (as in the intensive distribution system). Only one agent might be necessary in a large region for specialty products; therefore, an exclusive distribution system should be considered.

After selecting the appropriate type of distribution, many other decisions must be resolved. Efficiency of the distribution center must be optimized. Where (or if) to build warehouses is sometimes the first consideration. Geographical location is influenced by accessibility (highway, railroad, airport). Inventory taxes levied by certain states is another factor. States and regions compete in trying to convince manufacturers to build centers in their area. Innovations in the warehouse system itself have provided great financial savings, however. Computer systems have been installed to help locate the area in which the product is stored and to notify inventory control when quantity of the product is low (*Transportation and Distribution* 1988).

Product dimensions and composition are concerns of distributing agents in a firm, because customers are now demanding faster delivery. The more cumbersome the item, the more cost the company will incur. However, today's delivery services can accommodate nearly any company's need.

Many companies are finding that distribution services are economically more feasible than creating their own center. Inventory storage, distribution, and reports offered by Federal Express, for example, are attractive to companies. Distributing products from centers instead of the manufacturing plant makes American companies more competitive in the world market. Delivering to the carrier storage facility is now part of the just-in-time concept, reducing inventory costs (Bowman 1988).

Product distribution is a major marketing element considered by the company. The distribution method influences the products they are willing to produce, and their choice of customers to serve. Whether an intensive (many agents carry the product), selective (a few key distributors carry the product), or exclusive (one agent in a specific area carries the product) delivery method is used, the targeted market segment must be specifically defined to determine how it can best be reached and satisfied.

Service

Inquire: How frequently will this product require service?
How expensive is that service likely to be?
Will costs be associated with "down-time"?

Is this a product that is likely to require little or much after-the-sale service? Is it prone to repair/breakdown? Microcomputers, for example, do not usually require a great deal of service, and component reliability is usually a selling point. When they do need it, however, it can often be expensive. Consider whether the service staff needs to be specially trained, or if the repair/service is likely to be quite expensive or require considerable down-time. Does a service system already exist, and does this pose a competitive threat?

GM, Ford, and Chrysler are all touting their own versions of the "Buyer Protection Plan" to take the high cost out of repairs. IBM has its own maintenance agreements and competes with numerous local companies for the opportunity to service the product after the sale. Technical skill is important in all of these areas, however. Not all products require follow-up service. Door hardware, light fixtures, furniture, etc., are all examples of products that require little or no follow-up service.

If consumers purchase an expensive product, they typically will try to have it repaired before purchasing another. Post-purchase service is an important factor to consumers. If they expect efficient, inexpensive service, consumers will perceive less risk in owning the product. However, if the product requires extensive, special servicing, it may not penetrate the market as quickly as it would if it were less complex.

Videocassette recorders (VCRs) were in at least 50 percent of American homes in 1987 according to *Newsweek.* More recent statistics have shown penetration of up to 70 percent through 1991. Sales of VCRs have increased as prices have declined. Unfortunately, consumer repair costs or the loss of use due to occasional breakdowns are not always considered at the time of purchase.

VCRs are complex machines consisting of relatively reliable

electronic parts and somewhat less reliable mechanical parts. Motors, belts, and gears can wear quickly under constant use. These machines require more care than a TV or stereo. Dust and extreme temperatures affect component parts. Cheap, low quality tapes can also clog the recording heads.

Most problems stem from owner misuse. Consumers try to force tapes into the machine without turning on the power, which can damage the tape loading mechanisms. Service people report that plastic toys, fruit, and nests of insects have understandably jammed some VCRs. Consumers who purchase VCRs with many features should become even more concerned with VCR care.

In purchasing a highly featured and complex VCR, one must realize that the repair will also be more complex than for a basic model. Repair cost and service time are typically less for a simpler model. Though the basic model can usually be repaired more cheaply, some consumers opt to replace it with a new one instead of having it serviced (Pauly, Tsiantar, and Anderson 1987).

A product that requires very little service will be viewed more positively by consumers. For example, a compact disc player, when properly maintained, will last for years: there are no major frictional components; the music is created with laser-pick-up instead of needle/vinyl contact; the machine doesn't have as many mechanical components as the VCR; the CD player features are more electronically based; CD players are much easier to be serviced; and, virtually every major stereo merchandiser has an adequate repair department.

High service requirements usually inhibit market penetration; however, the VCR industry is flourishing. Consumers apparently feel VCRs are worth the difficulty and risk of maintenance, enjoying the convenience of viewing and taping movies in their own homes.

Competitive Factors

Competition, it is said, drives business to greater heights. Thousands of new products will be introduced this year, and only a few will survive. Competition is of two types, *product* and *brand.* Product competition occurs when one product is substituted for another to do the same job. I may not have a bar of hand soap, so I use liquid soap instead. Similarly, I may be out of toothpaste, but because I have a speaking engagement, I choose to use a breath-freshening gum.

Brand competition exists when two essentially alike products from different manufacturers compete for the same customers along similar dimensions of performance. Dial brand soap may compete with Safeguard brand soap in the hand-soap market. The Plymouth Voyager competes in essentially the same product-market with the Dodge Caravan.

In this section we examine dimensions of competition, focusing on those that may be highly determinant of a product's ability to survive in the jungle we call the marketplace.

Appearance

Inquire: In comparison with the competition in this particular product market, how "good-looking" does this product appear?

This criterion relates to the overall appeal of the product based on its physical characteristics. Is it attractive, ugly, pleasant, etc.? In total, how does it appear vis-à-vis the competition? Crown Royal brand whiskey comes in a glass container of unusual shape and obviously expensive production quality–accentuated by being packaged in a velvet sack. Generic brands of whiskey come in clear, simply rounded glass containers engineered for production simplicity. Labeling is simple, and packaging is plain.

The "clear" trend has entered the marketplace. Clear has been associated with healthier and more environmentally correct products. Palmolive Sensitive Skin is a clear product because it does not contain any dyes or irritants. Crystal Pepsi may be adopting this philosophy.

"A lot of it is perception," says Sheldon E. Rysner, a partner in Chicago-based package and design firm Murrie Lienhardt Rysner & Associates. "Water is pure and clear. It has only two elements, whereas a lot of these things are made with a lot of stuff and making them clear implies that there are less things in it" (Miller 1993).

Procter and Gamble's Ivory Liquid dish soap has adopted the clear image. The company says it made its product clear to further emphasize its grease-cutting ability. P & G spokeswoman Lynn Hailey said, "There had not been any changes in the product for quite a while, and making it clear was a way to signal in a dramatic way that the product had been changed" (Miller 1993). Since the product's release in the spring of 1992, sales have noticeably improved.

The clear fad is even reaching the brewing industry. Coors is test marketing Zima Clear Malt, a transparent, alcoholic beverage. Should this product's appearance prove successful, Coors may release the product for mass consumption during the summer of 1993.

Current trends in transparent products and packages may be a rage which could terminate its own intrinsic value. If too many products become colorless, the "clear" product representation will lose its credibility as a purer and more natural product (Miller 1993).

Function

Inquire: Does the product perform as claimed?
How does the performance compare with similar substitute products?
Specifically, what are the advantages and disadvantages compared with the competition?

What does it do? How well does it perform? What are its basic performance attributes? These questions are designed to get you thinking about the functional advantages and disadvantages in comparison to competitors' products.

The market for hand-held cordless telephones is growing. Many make the claim that they are noiseless and that they have an effective range of 100 feet. Does the innovation compare well or poorly with the competition? As for soft drinks, Coke has touted that you will drink Diet Coke "just for the taste of it." Does this ring true? Does it actually taste better?

A product's function fulfills a need. Whether the function is basic (actual activity or purpose of the product design) or aesthetic (appearance or status of the product), it satisfies the consumer through consumption. Evaluating a product's function involves comparison with similar products as well as analysis of the extent to which it serves its basic and aesthetic functions.

Black and Decker's Dustbuster hand-held vacuum cleaner fulfills the consumer's need to clean small household messes without bringing out the heavy artillery (Hoover upright). The basic function is to provide a convenient way to clean small domestic disasters.

The Dustbuster was introduced in 1979. It contains three nickel-cadmium rechargeable batteries. At under two pounds, the hand-held vacuum can be carried to the mess, switched on, and placed back in its holder before one could get the plug in the outlet using a conventional vacuum. Consumers can use the mini-vacuum to pick up potting soil, hair, and cigarette butts, and it can be used on car interiors to name but a few uses.

Aesthetically, this vacuum may leave something to be desired. However, it is rare that someone would hang it next to their Stradivarius. The vacuum is composed of a sturdy, shock-resistant plastic, snout-like body about 14 inches long and four inches high, and comes in several colors.

Since the Dustbuster's introduction, competitors such as Norelco, Eureka, Douglas, Sanyo, Sears, and Conair have tried to capture a portion of the mini-vac market. The Dustbuster-Plus, however, was found to be the superior mini-vac according to 1991 *Consumer Reports* product reviews. The only disadvantage found by the reporter was that the charges only lasted approximately ten minutes, a longstanding complaint (*Consumer Reports* January, 1986).

Overall, the mini-vac serves an (albeit somewhat induced) important basic function for consumers. Its only competitor at present is the cumbersome standard vacuum. Convenience of operation is its strong point. The Dustbuster appears to be the most visible, and consequently is the top seller in fulfilling the "small mess clean up" need.

Durability

Inquire: How long is the life of the product in comparison with the competition?
How well will the product withstand normal (or exten-

sive) wear and tear?
Would durability be a major brand strength or weakness?

Durability relates to buyers' perceptions of the product's ability to withstand use. While there is always a bit of price-durability trade-off, keep in mind that price is a separate issue. Consider durability as a distinct factor, namely how the product will stand the tests of use and time. The DieHard line of Sears batteries is known for durability; so are EverReady brand household batteries. Certain lines of Monroe brand shock absorbers are warranted to last the lifetime of your car.

In a competitive product market it is evident that the manufacturer offering the product with the best performance at a relatively low price will obtain the greatest market share. In the case of auto polishes, durability and shine are the desirable performance characteristics. Since many competitors exist, many similar products are offered.

In order for manufacturers to attract customers, they must differentiate their product from substitute products. Advertising or packaging can appeal to the novice buyer of car polishes, but for the veterans durability and price are the major factors.

The most important criterion in the car polish market is durability. A polish's long-term ability to protect the automobile from environmental elements such as dirt, salt, road oil, and rain demonstrates the effectiveness of the product. The time span over which the polish is able to bond to the car surface determines the product's durability. Durability is desirable because once the polish is worn away, it cannot protect the finish.

Consumer Reports challenged the product durability claims of various auto polishes. Interestingly, one product that claimed to provide 12 months protection ("Westley's Concentrate") lost its water repellent ability after a single aging and washing cycle. According to *Consumer Reports,* the durability of the product was overstated. Five brands of polishes were able to effectively seal the surface through six cycles, with the most economical and effective of the five being Nu Finish liquid (*Consumer Reports* June, 1988).

To effectively sell an auto polish the manufacturer needs to convince the consumer that its product is not only reasonably priced, but its durability level is appropriate. However, the manufacturer should not exaggerate its product's ability. If the product does not perform as the consumer expects, it will probably not be repurchased.

Auto batteries are products where chief consideration is given to durability. Sears' DieHard brand battery is depicted in advertisements starting a car in sub-zero conditions and in blistering desert heat, as well as starting an over-the-road semi-truck. Sears is trying to show consumers the power and durability of its product.

Progress in the battery market often originates with the U.S. Department of Energy, or Electric Power Research Institute laboratories. The latest advancements are in the battery's "toughness" and storage capabilities.

Energy storage is being improved with further developments in the sodium/sulfur batteries. A system using sodium/sulfur, instead of the conventional lead-acid, can store three to four times more energy per unit weight. Stabilization of the battery is also being developed by using a lithium/iron disulfide cell (Rawls 1985).

Progress in the durability of the battery will improve market appeal. Consumers may soon be able to purchase car batteries with these qualities, adding to the value of their purchase.

Price

Inquire: Will the proposed price of this product be a competitive advantage in the targeted markets?
What are the objectives to be achieved by the chosen price-level?

Many analysts consider this to be among the most important of the variables in the marketing mix. Accordingly, pricing decisions are always made with target markets in mind, with the knowledge that a range of acceptable prices exists for the product within those target markets. Other considerations that are taken into account include competitors' prices, the impact of price on the demand curve, and the degree of pricing flexibility that the initial price provides the marketer in meeting financial goals. Here, consider price relative to that of the competition: will it be higher or lower?

Take into account factors such as production, distribution, and promotional costs.

Chivas Regal brand Scotch whiskey is comparatively quite expensive, but not nearly as expensive as some of the straight malt scotches such as Glenfiddich brand. Alternatively, many scotches can be purchased for less than ten dollars per bottle. Price for each is apparently appropriate for the markets they target, each more or less price-sensitive.

Generally, the law of demand states that when the price of a product or service decreases, the quantity demanded will increase. If the price of a product is lower than that of substitute products, then consumers will more than likely prefer the cheaper product.

Telephone modem technology has advanced quickly. These improvements have caused the price of modems to decrease while their efficiency has increased. More and more computer users are acquiring modems due to this combined incentive.

Market competition forced computer modem prices down. Many purchasers who were hesitant to link computers began to consider the possibility. Company mergers are making it necessary for different plants to share information, and the modem is an ideal way to fulfill this need. The lower price has resulted in vast modem sales (Seymour 1987).

The question of what price to set for a new product always arises. The complex answer is "whatever people are willing to pay for it." An innovative approach was used by the Noumenon Corporation.

Noumenon introduced a software package, Intuit, into the computer market at a price of 395 dollars. The package would not sell. Noumenon decided to apply economic theory to its pricing policy.

Noumenon reintroduced the product at a cost of 50 dollars and increased the price by 20 dollars each week until sales began to decline. The data collected was plotted to determine the intersection of demand and supply. It was determined that sales were optimal when Intuit sold for 90 dollars. Noumenon soon captured the price-conscious consumer market for its particular software (Walker 1985).

In a highly competitive market, it is often necessary to use price differentiation when setting pricing policy. In the airline industry, for example, discounts are offered for people who book well in advance of their flight date. The airline can predict the number of meals or the amount of empty seats and use this information to lower its real costs.

Consequently, spontaneous fliers such as business travelers are burdened with higher prices. Airlines know that these people are willing to pay more for an airline ticket than college students are. The college student will typically book in advance to receive a discount, whereas the business flier will pay for the convenience of booking a flight with little or no advance notice (Schwieterman 1985).

Setting a price is a complicated process that requires studying the potential consumers. Market research is often the most critical element in the decision. Not only must the cost of producing a product or service be considered, but the attitude of the consumer toward the price must be evaluated before a pricing policy can be established.

Existing Competition

Inquire: Who is the competition?
What are their business strengths?
What are their business weaknesses?

How strong are your competitors? How much of the existing market share do they have? Do the existing competitors enjoy any barriers to new entrants?

It is often argued that the oil companies have monopolized the oil industry in this country, yet several dozen very large companies compete for gasoline business alone. However, the financial assets, technological know-how, established distribution channels, etc., all serve as competitive barriers to any new entrants.

Only the strong survive in America's competitive markets. A new company has to compete on at least the same level as competitors, or the targeted segment may not be swayed. This is especially evident in the brewing industry. Existing competition in the market is so intense that it is virtually impossible for a new company to enter the market on a national level.

The beer market is dominated by Anheuser Busch and Miller, making it difficult for Coors and Stroh Brewing Company–though still major players–to obtain substantial market shares. The competition have all but cornered major sporting events and their television viewers. It could be argued that one of the most intimidating barriers to entering this market is the level of promotion that products receive.

The concept of capitalizing on small sponsorship as opposed to major sponsorships has become a viable option for smaller brewers because of the expense involved in national television advertising. People are more likely to see a Coors or Old Milwaukee banner or logo on walls or cars rather than as a commercial during the superbowl. The difference is a strategic one: smaller brewers with less popular products are decentralizing and localizing national promotions, while Budweiser, Miller Lite, Bud Light, and Coors Light continue to use a national approach.

The consequence of the advertising strategies are predictable. Budweiser is the top seller for the beer industry at (23.2 percent) with Miller Lite at (9.7 percent), Bud Light at (7.2 percent), Coors Light at (6.7 percent) and Busch at (5.2 percent) rounding out the top five. Strategy is important when trying to influence a market where existing competition is intense. In the beer market, where existing brands may pose otherwise insurmountable barriers (e.g., advertising intensity or excellent distribution), it is critical to innovate in all areas (Teinowitz 1993).

The electric razor market is another highly competitive arena. The three top players in men and womens' electric shavers, Remington Products, Panasonic Company, and Norelco Consumer Products Company, must not only compete with one another but must try to lure "wet" razor users to their products.

Existing competition is fierce, especially around the holidays. Remington, Norelco and Panasonic all increased their advertising budgets for the fourth quarter of 1992 because roughly one-half of yearly sales result from gifts purchased during the holiday season.

All three companies are standing by their familiar and still effective advertising campaigns. Remington's Victor Kiam is touting his familiar adage, "Shaves as close as a blade or your money back." Norelco plans to re-release its commercial featuring Santa riding across the snow atop its triple head razor. Panasonic is combining its "Smooth Operator" campaign such that it addresses both men and women's shavers (Davis, 1992).

New Competition

Inquire: How long will it take for the competition to develop and enter the market with a competitive brand? A competitive product?

This question asks you to make a guess concerning the lead your innovation may temporarily enjoy in the market. Make an informed guess, however, and be able to show *why* you think new competitors will enter, and *how soon*. What will *their* "hook" into the market be? Take into account such factors as the rate of innovation in the industry, the degree of risk associated with entry, the barriers the innovation may have to its advantage, etc., and draw an educated conclusion. Research is very important here.

The company or companies introducing the technological innovation that is likely to replace audio compact discs (CDs), will, for a period of time, enjoy the barriers of learning costs, capital investments, etc. But, recall that CDs were introduced just a few short years ago. Technology changes very rapidly in electronics product markets.

When an invention proves to be successful, it is not long before "copy-companies" produce similar products. Success has a legion of followers. New competitors may enter with a similar product at a cheaper price or with additional options. Xerox may have produced the first copier, but Ricoh, Sharp, and Canon were soon in the same market.

Block Drug Company's Sensodyne and Proctor and Gamble Company's Denquel have all but monopolized the sensitive toothpaste market. But with the average age of the population rising, the market sales potential is becoming attractive to new competitors.

Sensitive toothpaste is used by individuals who experience pain in their gums because of hot, cold, or sweetness. Teeth become more sensitive as gums recede later in life. Approximately 18 percent of the adult population now experiences this condition and this percentage is expected to increase over the next two decades.

This inevitable growth has prompted SmithKline Beecham to

introduce its AquaFresh Sensitive toothpaste. Sensitive toothpaste is almost identical to common toothpaste with the exception of adding potassium nitrate. Although potassium nitrate relieves the pain, it has a distinct and rather unpleasant taste. Disguising this taste is the key to success in this market.

AquaFresh plans to penetrate the market by underpricing the existing competition. Where Sensodyne and Denquel sell for nearly twice the price of common toothpaste, SmithKline Beecham plans to release its AquaFresh Sensitive for approximately 30 percent more than its common toothpaste (Winters 1992). Even in a market as mature and crowded as this, innovation is important. The established brands are now under seige.

Protection

Inquire: Is the device protected by one or more patents?
Is the technology obvious, could anybody do this, or is it genuinely an innovation?
Of the types of protection available, which would seem to offer the best form of property protection?

Simply put, this factor relates to the extent to which the product or brand can be insulated or protected from the competition. How many people do you know who can say with certainty that they know the formula for KFC's "unique blend of 11 herbs and spices"? How about the recipe for Coca Cola classic? Have you seen it in Rolling Stone? No! It's locked in a vault at Coca Cola headquarters. The most common ways of prolonging protection and preserving technological advantages are to obtain patents and copyrights, use trademarks, and to preserve trade secrets. Both Coca Cola and KFC have chosen to rely heavily on the trade secret approach.

Protecting the ownership of an idea is the concept behind the patent. Invention is the major factor in the continuing growth of technology, so intellectual property must be protected. Court decisions on infringement of ownership of an idea are prevalent. This is due to increasing favor toward those people who have exercised their patent rights. People, nations, even the U.S. government have used other persons' ideas without originator consent, and have been convicted of infringements.

A recent success story is that of Texas Instruments. TI has a patent for its DRAM (dynamic random access memory) chips. Japanese firms had been making the chips under a limited license, dominating the market. TI felt that the Japanese were taking advantage of the license, so when the next negotiation occurred, TI increased the license cost. Companies that refused to recognize TI's new demands and continued to make the chips were taken to court. Nine Asian manufacturers were forced to submit to new licensing agreements that cost *five to ten times more* than before.

Since this court battle, Polaroid has sued Eastman Kodak for infringements on its instant camera patents; General Motor's Hughes Electronics Corporation has sued the U.S. government for satellite technology infringement, and Apple sued Hewlett-Packard and Microsoft for copying its software style. The patent supporters insist that patents protect ideas and encourage innovation. Those opposed to strong patent rights contend that they create monopolies.

Patents are an effective means of protecting ideas, if the courts enforce the inventors' rights. However, in some Asian countries many manufacturers are capitalizing on developed products that are selling well. They simply copy the product and sell it. This method is attractive because the development costs are avoided, and only proven high volume, successful products are manufactured. The U.S. International Trade Commission (ITC) has been created to relieve or possibly eliminate the overseas copying problem, and it has seen some success. For example, Bell, a motorcycle helmet manufacturer, used the ITC to keep Japanese companies from copying their ventilated helmet without a license. Innovators should be aware of the actions that can be taken against the infringing companies. But more important, perhaps, is the recognition that simply obtaining protection (whether through a patent or other means) is not enough. The process bears no relationship to the interests consumers may have. Also, the holder must be ready, willing, and able (financially) to protect that investment. Seeking opinions of a registered patent agent or attorney is a must before any product is actually introduced, but, obviously, it is just one of many factors that need attention (Alster 1988).

SUMMARY

This chapter introduced a number of issues relating to the use of screening models and then detailed the factors used by the PIES-II tool. The use of this or any other evaluation instrument has certain strengths and weaknesses. Some weaknesses relate to the stage of the process, others are derived from instrumentation, and still others arise from judgment errors. However, the use of a comprehensive and systematic set of procedures in screening will help to minimize losses from poor candidates, and assist in focusing efforts on a smaller pool of more promising prospects.

REFERENCES

Advertising Age. "Coke and Pepsi Carve a Tiny Niche in Huge Market." June 9, 1986, Vol. 51, p. 56+.

Advertising Age. "Party on Wayne: New Products 1992." December 21, 1992, p. 12.

Advertising Age. "Total U.S. Ad Spending by Category and Media." September 23, 1992, p. 8.

Albers, S. and K. Brockoff. "Optimal Product Attributes in Single Choice Models." *Journal of the Operations Research Society*, Vol. 31, 1980, pp. 647-55.

Alster, N. "New Profits From Patents." *Fortune*, April 25, 1988, pp. 185-90.

Baker, K. "The Comparative Analysis of Models for Use in New Product Screening Decisions." PhD Dissertation, University of Oregon, 1980.

Baker, K. and G.S. Albaum. "Modeling New Product Screening Decisions." *Journal of Product Evaluation Management,* Vol. 1, 1986, pp. 32-35.

Bowman, R. "Pumping Parts Through the Distribution Pipeline." *Distribution*, September 1988, pp. 78-84.

Calantone, R. and R.G. Cooper. "New Product Scenarios: Prospects for Success." *Journal of Marketing*, Vol. 45, (Spring) 1981, pp. 48-60.

Colombin, L. and L. Jourdaine. "Innovation is the Future of Glass in the Automobile." *Glass Industry*, July 1988, p. 24.

Consumer Reports. "Auto Polishes: They Promise Dazzle and Durability. What Do They Deliver?" June 1988, pp. 377-79.

Consumer Reports. "Dry-Cell Batteries." November 1987, pp. 703-706.

Consumer Reports. "Follow-up: Some Suzuki Rollovers." January 1988, p. 487.

Consumer Reports. "Hand-Held Cordless Vacuum Cleaners." January 1986, pp. 21-23.

Consumer Reports. "Sunscreens." June 1988, pp. 370-74.

Cooper, R.G. "The Dimensions of Industrial New Product Success and Failure," *Journal of Marketing*, Vol. 43, 1979, pp. 93-103.

________. "Selecting Winning New Product Projects: Using the NewProd System." *Journal of Product Innovation Management*, Vol. 2, 1985, pp. 34-44.

Cooper, R.G. and U. de Brentani. "Criteria for Screening New Industrial Products." *Industrial Marketing Management*, Vol. 13, 1984, pp. 149-56.

Cortez, John P. "New Direction for Domino's." *Advertising Age*, January 4, 1993, p. 3+.

Crocker, O.L. and R. Guelker. "The Effects of Robotics on the Workplace." *Personnel*, September 1988, pp. 26-36.

Davis, R. "Electric Razors Plan Aggressive Fourth Quarter." *Advertising Age*, October 12, 1992, p. 20.

Engel, J., R. Blackwell and D. Kollat. *Consumer Behavior*, 3rd Ed. Hinsdale, IL: Dryden Press, 1978.

Fisher, A. "Science Newsfront." *Popular Science*, October 1988, pp. 8-9.

Fisher, C. "Hershey, Mars, Nestlé Offer Cheer." *Advertising Age*, December 21, 1992, p. 27.

Flower and Garden. "Snow Removal Equipment." December/January 1987, pp. 74-75.

Giannetti, T., W.T. Jones, A.G. Feliu, K. Ross, W.B. Comegys and L. Maloney (Chief Editor). "Fresh Answers to Legal Worries." *Design News*, April 18, 1988, pp. 66-72.

Glaberson, W.B. "Did Searle Close Its Eyes to a Health Hazard?" *Business Week*, October 14, 1985, pp. 120-22.

Goldberg, N. "Beyond Autofocus: Where Should Technology Take Automation Next?" *Popular Photography*, August 1981, Vol. 88, p. 66+.

Gray, R. "Add Exposure, Make 'em Official." *Ward's Auto World*, May 1986, pp. 81-82.

Hampton, W. "Two-Stroke Engines Have Detroit Buzzing." *Business Week*, January 25, 1988, p. 102+.

Hawkins, D.I., R.J. Best and K.A. Coney. *Consumer Behavior: Implications for Marketing Strategy*, 3rd Ed. Plano, TX: Business Publications, Inc., 1983.

Jacobson, M.F. "Let People Know What They're Eating." *The New York Times*, January 12, 1986.

Labich, K. "The Innovators." *Fortune*, June 6, 1988, pp. 50-64.

Lawrence, J. "Rivals Readying First Challenges to Huggies Pull-Ups." *Advertising Age*, October 12, 1992, p. 4.

Marketing News. "Beer Sales Decline." February 17, 1992, p. 1.

Marketing News. "You Can't Sell From Afar." February 29, 1988, p. 20.

Miller, C. "Trendy Marketers Want Consumers to See Right Through Their Products." *Marketing News*, February 1, 1993, pp. 1-2.

________. "Little Relief Seen for New Product Failure Rate." *Marketing News*, June 21, 1993, pp. 1t.

Minicucci, R. "Training Non-Traditional Users." *Today's Office*, March 1987, pp. 51-56.

Mowen, J.C. *Consumer Behavior*. New York: Macmillan, 1987.

Murphy, P.E. and B.M. Enis. "Channel of Distribution Concepts: Channel Design

Decisions." *Marketing*. Glenview, IL: Scott, Foresman and Company, 1985, pp. 363-67.

Neal, L.A. "Taking on Environmental Hazards." *Real Estate Today*, March 1987, pp. 25-27.

O'Meara, J.T. Jr. "Selecting Profitable Products." *Harvard Business Review*, Jan.-Feb., 1961, pp. 110-18.

Papanek, V.J. *Design for the Real World*. New York: van Nostrand Reinhold Company, 1984, pp. 224-37.

Pauly, D., D. Tsiantar and M. Anderson. "Is Your VCR in the Shop? You Are Not Alone." *Newsweek*, October 19, 1987, pp. 62-63.

Perrault, W.D. and F. Russ. "Comparing Multiattribute Evaluation Process Models." *Behavior Science*, Vol. 22, 1977, pp. 423-31.

Rawls, R.L. "Material Developments Highlight Progress in Batteries." *Chemical and Engineering News*, December 16, 1985, Vol. 63, pp. 23-27.

Richman, L.S. "Why Throw Money at Asbestos?" *Fortune*, June 6, 1988, pp. 155-70.

Rosenthal, T.M. "Market Research Legwork Pays Off For Parking Meter Manufacturer." *Global Trade*, January 1988, pp. 8-9.

Schwedelson, R. "New Wave Data Base." *Direct Marketing*, March 1988, pp. 40-41+.

Schwieterman, J. "Understanding Airline Price Discrimination." *Consumers' Research Magazine*, October 1985, pp. 16-19.

Seymour, J. "Modem Prices Plummet as Speeds Skyrocket." *Today's Office*, June 1987, pp. 12-16.

Shocker, A.D., D. Gensch and L. Simon. "Toward the Improvement of New Product Search and Screening." *Proceedings American Marketing Association*, Fall 1969, pp. 168-75.

Smith, A. "Upjohn's Bald Ambitions." *Esquire*, October 1986, pp. 73-74.

Stahl, M.J. and T. Zimmer. "Modelling Product Development Decision Policies of Managers and Management Students: Differences Between Subjective and Relative Weights." *IEEE Transactions on Engineering Management* (EM-30), #1, 1983, pp. 18-24.

Stults, K. "Tarnish on the Golden Arches." *Business and Society Review*, Fall 1987, pp. 18-21.

Swasy, A. "Firms Grow More Cautious About New-Product Plans." *Wall Street Journal*, March 9, 1989, p. B17.

Teinowitz, I. "A-B Gains in Beer Share, Thanks to Hot Bud Light." *Advertising Age*, January 4, 1993, p. 2.

Transportation & Distribution. "Inside Warehousing and Distribution." March 1988, pp. 61-76.

Udell, G.G. and K.G. Baker. *PIES-II Manual for Innovation Evaluation*. Whitewater, WI: Small Business Development Center, 1980.

Walker, K. "Software Economics 101." *Forbes*, January 28, 1985, p. 88.

Wall Street Journal. "Avon Recalls 350,000 Bumble Babee Plush Toys." November 24, 1987, p. 32, col. 3.

Wall Street Journal. "Quaker Oats & Fisher Price Unit Pay $2.3 Million for 'Little People'." May 15, 1988, p. 30, col. 6.

Winters, P. "AquaFresh Gets Sensitive About Baby Boomer Teeth." *Advertising Age*, October 12, 1992, p. 4.

________ . "Industry Sweet on Flavored Schnapps." *Advertising Age*, July 21, 1986, pp. S4-5.

Chapter 3

An Information Support System for New Product Screening Decisions

INTRODUCTION

The information upon which the new product judgments are based is critical to new product screening. The adage "garbage in, garbage out" is a frank reminder that the decisions made by marketing and product managers can be no better than the insights and judgments upon which they rest.

This chapter provides the entrepreneur with an overview of marketing information systems (MKIS), and then focuses on a support-system template for gathering secondary information as an aid to product screening. The latter is particularly important to the novice or person not connected with a firm large enough to have an in-house research support staff. Unfortunately, and all too often, persons in this setting ultimately rely on visceral reactions in deciding the fate of new product ideas when, in fact, existing secondary information could clarify and improve decisions.

MARKETING INFORMATION SYSTEMS

A survey of Fortune 500 firms by McLeod and Rogers (1982) on uses of marketing information systems revealed that formal methods of gathering information are commonplace and that the information is used at a level where it can have the greatest impact, i.e., middle-level management, which has responsibility for creating and evaluating market-level programs. What is an MKIS, and what are the advantages and disadvantages of using a formal system for information collection and analysis?

Donald Cox and Robert Good (1967), original contributors to this field, (1967), define the concept as a "... set of procedures and methods for the regular, planned collection, analysis and presentation of information for use in making marketing decisions." Generally, this consists of a support system used to gather and process the data for decision making and an operating system that uses the data "... as an aid to planning and controlling marketing activities" (p. 145). In its most complete form, an MKIS is a network of structures fully integrated throughout the firm and meshed with the relevant outside organizations to ensure timely delivery of high quality insights that improve the marketing function. Data gathered in a continuous and balanced (for verification) fashion, relating to the needs of the organization, are the hallmark of the effort. One form such a system can assume is shown in Figure 3 (Tull and Hawkins 1987).

Depending upon the level of organizational commitment to the process, the applications and benefits of such an approach to management decision making can be many. Improved forecasts, quicker and incisive reactions to competitive thrusts, and better coordination with other organizational levels (e.g., finance, production) all represent advantages of decision making that relate directly to the marketing function.

It should be noted that not all MKIS efforts have been successful. One can easily drown in the sea of information that regularly pours in from a well-developed system. This problem, it seems, originates from two areas. First, the need to know does not always equate with the desire to know, and the consequences can be confusion and paralysis (Gilad and Gilad 1985).

Second, the presence of such a structure may lead to a tendency to rely too heavily on the (limited) sources of information the system represents. Systems analysts are quick to point out that the appearance of systematic and comprehensive structure can indeed be misleading.

Furthermore, it has been the experience of managers at General Electric that *even though* complex and sweeping support systems exist, many members who might benefit by their use will not participate (Lillis and McIvor 1985). Reasons are many, including lack of

FIGURE 3. Structure of a Marketing Information System

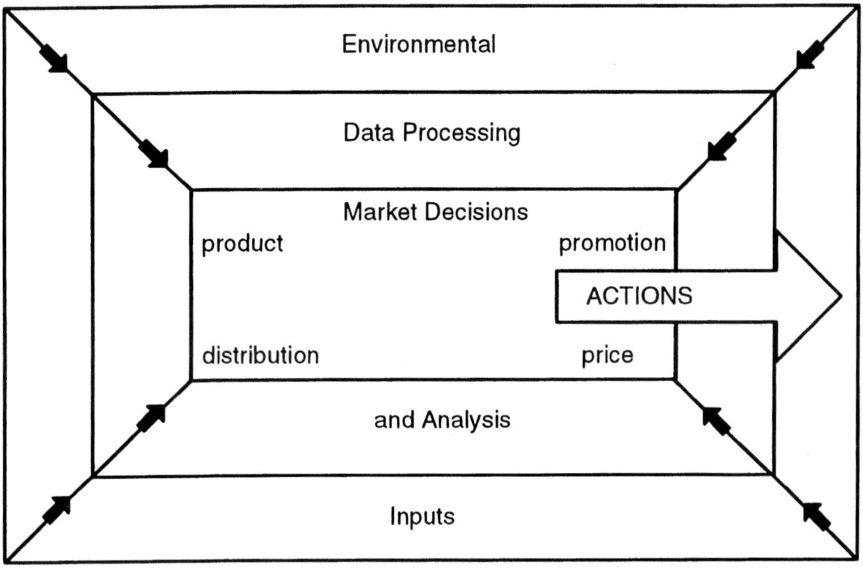

training, poor alignment of an otherwise generic system with existing marketing programs, or economic constraints.

Struse (1985), too, has been quick to point out that unless the system is well accepted by the social side of the organization, it is likely doomed.

The template proposed by Lesch and Hazeltine (1990) should be construed not as a comprehensive guide to research for new products or concepts, nor does it represent a complete MKIS. Rather, it structures the process of *secondary research,* i.e., organizes the review of existing and available information sources normally accessible through large, public libraries. The format is sufficiently general to be utilized for many aspects of marketing (new product introduction, competitor analysis, etc.), but, in concert with the criteria proposed for product screening, it lends structure to that routine marketing task. In combination with the PIES structure, our basic structure is complete.

SECONDARY DATA ANALYSIS

Libraries are frequently huge, almost foreboding structures, often of archaic design, and although this is not intentional, not always "user-friendly." The good news is that neither libraries nor librarians have been known to bite their patrons, and, except for an occasional visit by a supernatural or two (as in *Ghostbusters*), these buildings really are places to learn, and to learn to enjoy learning.

Unfortunately, business educators themselves do not always have the very best command of existing and growing collections that pertain to business decision making. This is a deficiency usually shared only among colleagues, but it is also reflected in the coverage provided to the topic of secondary research in business courses, and in marketing research classes in particular (Taschian, Pucely, and Crittendon 1985). Segal (1987) has shown through a regional survey that the marketing research industry is sorely dissatisfied with qualitative analytical skills of even MBAs.

Recently, a number of articles and texts of a broad nature have begun filling this gap. Stewart's (1984) monograph on secondary research has reached a wide and receptive audience, while bibliographies of business reference sources have been accumulating. An early and quick reference was provided by Goeldner and Dirks (1976) while the most comprehensive remains that of Daniells (1987). But, the critical questions of "how do I get started?" and "how do I know I found everything?" remain.

An excellent procedural response to these problems was promulgated by Milutinovich (1985) in which the author not only detailed common sources by topical headings, but suggested a macro fact-finding plan to guide the novice. A more complete search diagram was advanced by Emory (1980), and it is on the latter's efforts that the flow chart in Figure 4 is premised.

The chart is designed to be self-guiding, but brief instructions are in order. The following is a list of critical references.

TYPICAL SECONDARY RESOURCES

Periodicals	*Call Number*
Applied Science & Technology	Z7913.I7
Dissertation Abstracts	Z5055.U5A53

FIGURE 4

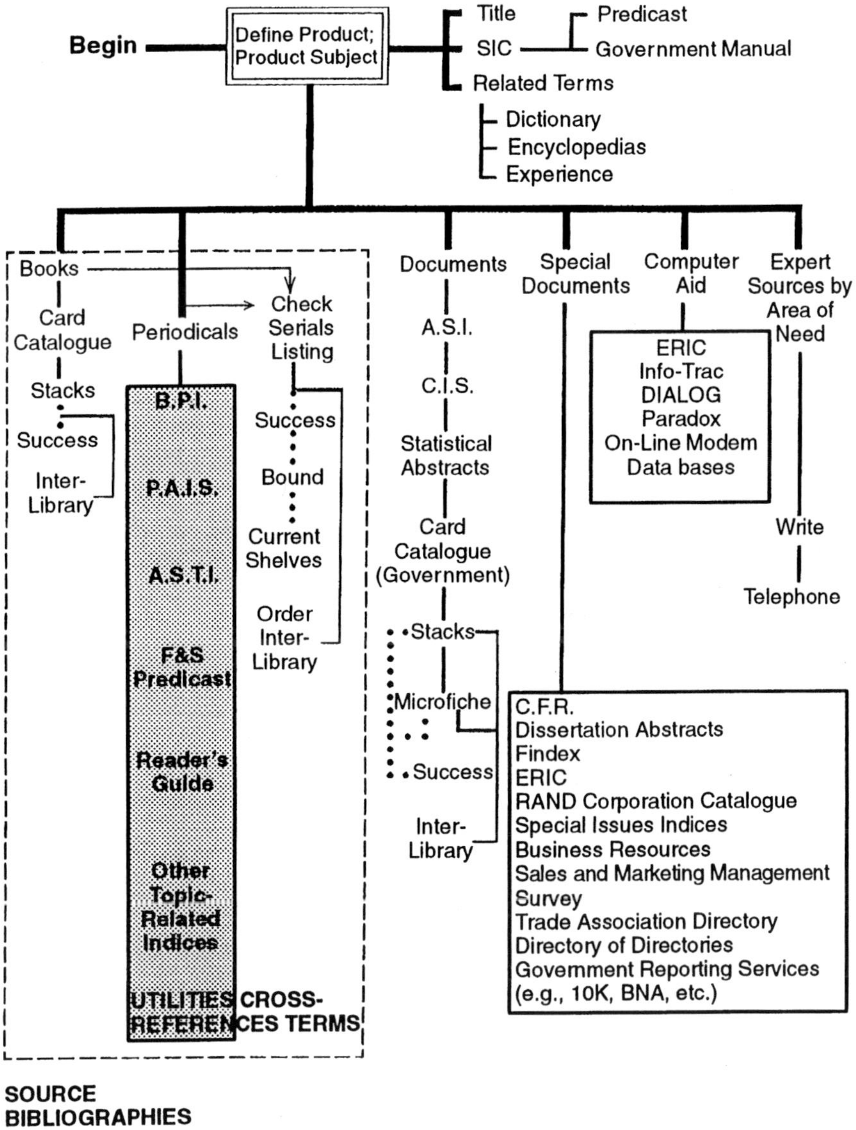
Begin
Define Product; Product Subject
Title
SIC
Related Terms
Predicast
Government Manual
Dictionary
Encyclopedias
Experience
Books
Card Catalogue
Stacks
Success
Inter-Library
Periodicals
Check Serials Listing
Success
Bound
Current Shelves
Order Inter-Library
B.P.I.
P.A.I.S.
A.S.T.I.
F&S Predicast
Reader's Guide
Other Topic-Related Indices
UTILITIES CROSS-REFERENCES TERMS
SOURCE BIBLIOGRAPHIES
Documents
A.S.I.
C.I.S.
Statistical Abstracts
Card Catalogue (Government)
Stacks
Microfiche
Success
Inter-Library
Special Documents
C.F.R.
Dissertation Abstracts
Findex
ERIC
RAND Corporation Catalogue
Special Issues Indices
Business Resources
Sales and Marketing Management Survey
Trade Association Directory
Directory of Directories
Government Reporting Services (e.g., 10K, BNA, etc.)
Computer Aid
ERIC
Info-Trac
DIALOG
Paradox
On-Line Modem
Data bases
Expert Sources by Area of Need
Write
Telephone

Periodicals	*Call Number*
Directory of Directories	AY2001.D52
Psychological Abstracts	BF1.P65
U.S. Census of Manufactures	DOC.C3.24
County Business Patterns	DOC.C3.204
Simmons Market Research Bureau	HF5815.U5S5421984
Leading National Advertisers	HF5801.A18
Value Line	HG4501.V26
Marketing & Media Decisions	HF5801.M38
Thomas Register	HE199.A2T4702
Forbes	HF5001.F6
Ad Forum	HF1040.A3
AdWeek	HF5801.A3401
Statistical Abstracts	HA202GLC
Advertising Age	HF5801.A276
Business Week	HF5001.B89
Business Marketing	HF5801.I4503
New York Times Index	A121.N45GLC
U.S. Industrial Outlook	HC101.U5402
Wall Street Journal	HG1.W26
Sales & Marketing Management	HF5438.A3401
Index Medicus	Z6660.I5
Social Sciences Index	A13.S6
Predicast	HG4961.F801
American Statistics Index	Z7554.U5A46
Moody's Industrial Manual	HG4961.M67
Standard & Poors	HG4057.A401
Standard Rates & Data	HF5905.S727
Index to Legal Periodicals Index	KF8.I5
Business Periodicals Index	Z7164.C81B983
Wall Street Journal Index	HG.W26
Engineering Index	Z5851.E62
Reader's Guide to Periodical Literature	A13.R48GLC
Public Affairs Information Service	Z7163.P9

In the case of the new product, a definition of what the item is should be determined. Generally, a working title for the research report contains a reference to this. The SIC (Standard Industrial Classification Code, a number) should be matched to the product category being researched. This can be ascertained either in the government manual (available from the librarian) or found in the *Predicast* index. It is always helpful to generate a list of related terms or synonyms for the product since not all indices will use the same words or terms in referencing.

Next, begin the book search, and when a systematic and (for the immediate time) comprehensive number of these has been identified, a preliminary analysis is in order.

What is comprehensive? How do you know your search has been systematic? The answer to both is to employ an *iterative* procedure. First, from your preliminary pass through the texts, did you find information necessary to answering your questions concerning the viability of your product along each of the 33 criteria? Second, was the data you gathered sufficient to adequately (authoritatively, conclusively) answer questions for each of the categories? In regards to the second issue, always consider the recency of the information and the extent to which the data *directly* (or only *partially*) addresses your questions.

If, as is the case in most "first passes," you find you cannot respond "yes" to both questions, you should proceed to "mine" or "source" (digging deeply into) the bibliographies of your texts, looking for other reference material.

Simultaneously, proceed to the indexes for the periodicals that might contain your needed data. Be sure to utilize and cross-reference your synonyms, and again, once you have located the articles, "mine" those sources to their end(s) to help fill any gaps.

Government documents are frequently the only sources of some information. Check those indexes (e.g., ASI and CIS) and remember that most libraries (even depositories) receive the majority of their documents on microfiche. You will not usually find a paper copy.

Identifying relevant special documents is often the key for tougher questions. Frequently, a trade association will have sponsored a study that may solve your problem, or an organization such as the RAND Corporation may have conducted a unique, but sufficiently general, study on the issue to be helpful.

Various computer-assisted on-line search techniques are available, some more expensive than others. We do not usually recommend attempting these until you understand the logic of the paper documents. If you try these without those competencies, your searches will be quite expensive in many cases because the researcher usually will not have narrowed search terms adequately, or will have to learn the search logic while on-line, at long distance rates.

Finally, contact with expert sources, local or national, can often clarify a technical area. Trade Associations and directories can help in identifying key experts, but your local university will often be your best first contact.

If you follow these procedures and correctly identify critical publicly available sources, it is likely that you will be able to address many, if not all, of the issues raised by the evaluation protocol. Your answers will vary in quality, of course, as a function of the data available and how well it fits with your needs. And finally, it is best to remember that even experts, when faced with the *same* information, may disagree.

SUMMARY

This chapter introduced the concept of an MKIS and secondary data decision support template. The procedures for using the system were discussed and a table of critical documents introduced. Your competency in the use of the template and documents will vary directly as a result of the amount of time spent in learning the source logics, the "true" nature of information available, and the accessibility of data/documents.

REFERENCES

Cox, D. and R. Good. "How to Build a Marketing Information System." *Harvard Business Review*, May-June 1967, pp. 145-54.

Daniells, L.M. *Business Reference Sources*. Cambridge, MA: Harvard University Press, 1987.

Emory, W. *Business Research Methods*. Homewood, IL: Irwin, 1980.

Gilad, B. and T. Gilad. "A Systems Approach for Business Intelligence." *Business Horizons*, Vol. 28, No. 5, 1985, pp. 65-70.

Goeldner, C.R. and L.M. Dirks. "Business Facts: Where to Find Them." *MSU Business Topics*, Summer 1976, pp. 23-36.

Lesch, W.C. and J. Hazeltine. "Secondary Research, New Product Screening, and the Marketing Research Course: An Experiment in Structured Decision Making." *Journal of Marketing Education*, Spring 1990, Vol. 12, No. 1, pp. 30-39.

Lillis, C. and B. McIvor. "MDSSs at General Electric: Implications for the 1990's from Experiences in the 1970's and 1980's." In *Marketing in an Electronic Age*, edited by R. Buzzell. Cambridge, MA: Harvard Business Press, 1985.

McLeod, R. Jr. and J. Rogers. "Marketing Information Systems: Uses in the Fortune 500." *California Management Review*, Vol. 25, No. 1, 1982, pp. 106-18.

Milutinovich, J. S. "Business Facts for Decision Makers: Where to Find Them." *Business Horizons*, Vol. 28, No. 2, 1985, pp. 63-80.

Segal, M. "Graduate Programs of Instruction in Marketing Research: A Need Assessment." *Journal of Marketing Education*, Spring 1987, pp. 3-11.

Stewart, D.W. *Secondary Research: Information Sources and Methods*. Beverly Hills: Sage, 1984.

Struse, R.W. "The Four P's of Marketing Decision Support Systems: Promises, Payoffs, Paradoxes, and Prognostications." In *Marketing in an Electronic Age*, edited by R. Buzzell. Cambridge, MA: Harvard Business Press, 1985.

Taschian, A., M. Pucely and V. Crittendon. "Marketing Research Texts: A Comparative View of the Area." In *1985 AMA Educator's Proceedings*, edited by R. Lusch, G. Ford, G. Frazier, R. Howell, C. Ingene, M. Reilly and R. Stampfl. Chicago: American Marketing Association, 1985.

Tull, D.S. and D. Hawkins. *Marketing Research*, 4th Ed. New York: Macmillan Co., 1987.

Appendix

File no. ___ Evaluator no. ___ Evaluator's name ____________

Group name ____________________

INNOVATION EVALUATION INSTRUMENT

Directions:

Circle the number of the response that best corresponds to your evaluation for each criterion. Be sure you answer all questions. Note that "don't know" and "not applicable" responses are coded "Unknown" and "NA." Be sure to use them when appropriate.

After the "Unknown" and "NA" responses, a space is provided for your working comments relative to that criteria. If you have any comments or explanations relative to your response, use this space.

On an attached sheet of paper please include specific information, comments, or suggestions relative to certain criteria that you wish to make to the client. These comments are highly useful in providing the inventor with additional information and insights.

Societal Factors

1. LEGALITY: In terms of applicable laws (particularly product liability), regulations, product standards, this idea/invention/new product . . .

1 – might not meet them even if changed
2 – might require substantial revision to meet them
3 – might require modest revision
4 – might require minor changes
5 – will meet them without any changes
6 – unknown
7 – not applicable

2. SAFETY: Considering potential hazards and side effects, the use might be . . .

1 – very unsafe, even when used as intended
2 – unsafe under reasonably foreseeable circumstances
3 – relatively safe for careful, instructed users
4 – safe when used as intended, with no foreseeable hazards
5 – very safe under all conditions, including misuse
6 – unknown
7 – not applicable

3. ENVIRONMENTAL IMPACT: In terms of pollution, litter, misuse of natural resources, etc., use might . . .

1 – violate environmental regulations and have dangerous environmental consequences
2 – have some negative effects on the environment
3 – have no effect on the environment if properly used
4 – have a slight favorable effect on the environment
5 – have a positive effect on the environment
6 – unknown
7 – not applicable

4. SOCIETAL IMPACT: In terms of the impact (benefit) upon the general welfare of society, use might . . .

1 – have substantial negative effect
2 – have some negative effect
3 – have no effect if properly used
4 – have some possible effects on society
5 – have considerable positive effects on society
6 – unknown
7 – not applicable

Business Risk Factors

5. FUNCTIONAL FEASIBILITY: In terms of intended functions, will it actually do what it is intended to do . . .

1 – the concept is not sound, cannot be made to work
2 – it won't work now, but might be modified
3 – it will work, but major changes might be needed
4 – it will work, but minor changes might be needed

5 – it will work–no changes necessary
6 – unknown
7 – not applicable

6. PRODUCTION FEASIBILITY: With regard to technical processes or equipment required for production, this invention might . . .
1 – be impossible to produce now or in the foreseeable future
2 – be very difficult to produce
3 – have some problems which can be overcome
4 – have only minor problems
5 – have no problems
6 – unknown
7 – not applicable

7. STAGE OF DEVELOPMENT: Based on the available information, there is . . .
1 – only an idea with drawings and/or description; no prototype
2 – a rough prototype which demonstrates the concept, but is not fully developed and tested
3 – a rough prototype with performance and safety testing completed
4 – a final prototype with testing completed; however, minor changes might be needed
5 – a market-ready prototype with testing completed
6 – unknown
7 – not applicable

8. INVESTMENT COSTS: The amount of capital and other costs necessary for advancement to the market-ready stage might be . . .
1 – greater than returns–investment may not be recoverable
2 – excessive–might not be recoverable
3 – heavy–probably recoverable
4 – moderate–should be recoverable
5 – low–should be recoverable within two years
6 – unknown
7 – not applicable

9. PAYBACK PERIOD: The expected payback time (time required to recover initial investment costs) is likely to be . . .

1 – over ten years
2 – seven to ten years
3 – four to six years
4 – one to three years
5 – less than one year
6 – unknown
7 – not applicable

10. PROFITABILITY: Profitability is defined as the extent to which anticipated revenue will cover the relevant costs (direct, indirect, and capital). Anticipated revenues . . .

1 – might not cover any of the relevant costs
2 – might cover direct costs but contribute minimally to indirect and capital costs
3 – might cover direct and indirect costs but might not meet capital costs
4 – might cover direct and indirect costs and meet minimum capital costs
5 – will cover direct and indirect costs and easily exceed capital cost
6 – unknown
7 – not applicable

11. MARKETING RESEARCH: The marketing research required to develop a market-ready product is estimated to be . . .

1 – extremely difficult and complex
2 – relatively difficult and complex
3 – moderately difficult
4 – relatively easy and simple
5 – very simple and straight-forward
6 – unknown
7 – not applicable

12. RESEARCH AND DEVELOPMENT: The research and development required to reach the production-ready stage might be . . .

1 – extremely difficult and complex
2 – relatively difficult and complex

3 – moderately difficult
4 – relatively easy and simple
5 – very simple and straightforward
6 – unknown
7 – not applicable

Demand Analysis Factors

13. POTENTIAL MARKET: The total market for products of this type might be . . .
1 – very small–very specialized or local in nature
2 – small–relatively specialized or regional in nature
3 – medium–limited national market
4 – large–possibly national market
5 – very large–national and possibly international market
6 – unknown
7 – not applicable

14. POTENTIAL SALES: Expected sales of this product might be . . .
1 – very small–economies of scale not possible
2 – small–economies of scale not likely
3 – medium–economies of scale possible
4 – large–economies of scale likely
5 – very large–economies of scale will be substantial
6 – unknown
7 – not applicable

15. TREND OF DEMAND: The market demand for products of this type appears to be . . .
1 – rapidly declining–product might soon become obsolete
2 – declining–potentially obsolete in the near future
3 – steady–demand expected to remain constant
4 – growing slowly–modest growth opportunity
5 – rapidly expanding–modest growth opportunity
6 – unknown
7 – not applicable

16. STABILITY OF DEMAND: The fluctuations in demand are likely to be . . .

1 – highly unstable–subject to severe unpredictable fluctuations
2 – unstable–susceptible to moderate unpredictable fluctuations
3 – predictable–variations can be foreseen with reasonable accuracy
4 – stable–modest variations can be accurately foreseen
5 – highly stable–not susceptible to fluctuations
6 – unknown
7 – not applicable

17. PRODUCT LIFE CYCLE: The product life cycle is likely to be . . .

1 – less than two years
2 – two to four years
3 – five to seven years
4 – eight to ten years
5 – more than ten years
6 – unknown
7 – not applicable

18. PRODUCT LINE POTENTIAL: The potential for additional products, multiple styles, qualities, price ranges, etc., is . . .

1 – very limited–single product only
2 – limited to minor modifications only
3 – moderate–multiple markets/use potential
4 – high–new product spin-offs likely
5 – very high–could be foundation of a new industry
6 – unknown
7 – not applicable

Market Acceptance Factors

19. COMPATIBILITY: Compatibility with existing attitudes and methods of use is . . .

1 – very low–impedes market acceptance
2 – low–some consumer conflict; will slow market acceptance

3 – moderate–no negative effects
4 – high–compatibility will aid marketing effort
5 – very high–will give market acceptance a strong boost
6 – unknown
7 – not applicable

20. LEARNING: The amount of learning required for correct use is . . .

1 – very high–expensive and/or time-consuming training required
2 – high–detailed instructions required
3 – moderate–normal instructions sufficient for most users
4 – low–minimal instructions needed
5 – very low–no instructions needed
6 – unknown
7 – not applicable

21. NEED: The level of need filled or utility provided by this innovation is . . .

1 – very low–gimmick soon forgotten by the owner
2 – low–would only superficially fulfill psychological non-essential needs
3 – moderate–fulfills both psychological and physical non-essential needs
4 – high–fulfills either basic psychological or physical needs
5 – very high–fulfills both psychological and physical needs
6 – unknown
7 – not applicable

22. DEPENDENCE: The degree to which the sale or use of this product is dependent upon other products, processes, or systems is . . .

1 – very high–lower rate of market penetration; higher costs
2 – high–low rate of market penetration; high costs
3 – moderate–reasonable market control and cost
4 – low–high rate of market penetration; low costs
5 – very low–higher rate of market penetration; lower costs
6 – unknown
7 – not applicable

23. VISIBILITY: The advantages and benefits are . . .
    1 – very obscure–very difficult and/or costly to communicate
    2 – obscure–requires substantial explanation
    3 – visible–requires some explanation
    4 – visible–easily communicated
    5 – very visible–advantages are obvious and easy to communicate
    6 – unknown
    7 – not applicable

24. PROMOTION: The costs and effort required to promote the advantages, features, and benefits are likely to be . . .
    1 – very high–prohibitive in relation to expected sales
    2 – high–relative to expected sales
    3 – moderate–in line with expected sales
    4 – low–relative to expected sales
    5 – very low–relative to expected sales
    6 – unknown
    7 – not applicable

25. DISTRIBUTION: The cost and difficulty of establishing distribution channels are likely to be . . .
    1 – very high–prohibitive in relation to expected sales
    2 – high–relative to expected sales
    3 – moderate–in line with expected sales
    4 – low–relative to expected sales
    5 – very low–relative to expected sales
    6 – unknown
    7 – not applicable

26. SERVICE: The cost and difficulty associated with providing product service is likely to be . . .
    1 – very high–will require frequent service and parts
    2 – high–will need periodic service and parts
    3 – moderate–will need occasional service and parts
    4 – low–need for service and parts will be infrequent
    5 – very low–will require little or no parts and service
    6 – unknown
    7 – not applicable

Competitive Factors

27. APPEARANCE: Relative to competition and/or substitutes, appearance is likely to be perceived as . . .

1 – very inferior–no customer appeal
2 – inferior–little customer appeal
3 – similar
4 – superior–has customer appeal
5 – very superior–has strong customer appeal
6 – unknown
7 – not applicable

28. FUNCTION: Relative to competing and/or substitute products, services, or processes, the function performed might be perceived as . . .

1 – very inferior–a significant disadvantage
2 – inferior–some competitive disadvantage
3 – similar
4 – superior–some competitive advantage
5 – very superior–a significant competitive advantage
6 – unknown
7 – not applicable

29. DURABILITY: Relative to competition and/or substitutes, durability of this product is likely to be perceived as . . .

1 – very inferior–a definite competitive disadvantage
2 – inferior–might be a competitive disadvantage
3 – similar
4 – superior–might be promoted as an improvement
5 – very superior–easily promoted as a major improvement
6 – unknown
7 – not applicable

30. PRICE: Relative to competition and/or substitute products, the selling price is likely to be . . .

1 – much higher–a definite competitive disadvantage
2 – higher–a competitive disadvantage
3 – similar
4 – lower–a competitive advantage

5 – much lower–a definite competitive advantage
6 – unknown
7 – not applicable

31. EXISTING COMPETITION: Existing competition for this innovation appears to be . . .
1 – very high–new entry might be difficult and/or relatively expensive
2 – high–new entry may be difficult
3 – moderate–market penetration can be gained with reasonable effort and expense
4 – low–a significant market share might be possible
5 – very low–entry might be easy and/or relatively inexpensive
6 – unknown
7 – not applicable

32. NEW COMPETITION: Competition from new entrants or competitive reaction is expected to be . . .
1 – very high–product lead will be very short
2 – high–product lead will be relatively short
3 – moderate–market share can be maintained
4 – low–product lead will be relatively long
5 – very low–a strong chance to sustain a large market share
6 – unknown
7 – not applicable

33. PROTECTION: Considering patents (or copyrights), technical difficulty, or secrecy, the prospects for protection appear to be . . .
1 – very poor–no legal protection or secrecy possible
2 – poor–no legal protection but some secrecy might be possible
3 – average–limited legal protection and some secrecy might be possible
4 – good–might be patented, copyrighted, or short-run secrecy possible
5 – very good–can definitely be patented, copyrighted, or long-term secrecy possible

6 – unknown
7 – not applicable

List in rank order the five criteria you believe to be most important relative to the commercial viability of this product/process.

criteria #	________	________	________	________	________
	1	2	3	4	5

Index